LUND STUDIES IN GEOGRAPHY

Ser. C. General, Mathematical and Regional Geography. No.14

AN ASSESSMENT OF WOODY BIOMASS, COMMUNITY FORESTS, LAND USE AND SOIL EROSION IN ETHIOPIA

A feasibility study on the use of remote sensing and GIS-analysis for planning purposes in developing countries

by

Author's address:

Remote Sensing Laboratory
Department of Physical Geography
University of Lund
Sölvegatan 13
S-223 62 Lund
Sweden

Tel. 046/108696
Telex: 33533 LUNIVER S

Lund University Press
Box 141
S-221 00 Lund
Sweden

Art nr 20057
ISBN 91-7966-016-9

Chartwell-Bratt Ltd.
Old Orchard, Bickley Road
Bromley, Kent BR1 2NE
England

ISBN 0-86238-158-4

Front cover:
Rainfall erosivity distribution of Ethiopia. The erosivity factor was derived as a function of the mean annual precipitation (Cf. Fig. 19).

Back cover:
The mean annual precipitation of Ethiopia. The plot was generated through a linear interpolation of the records of 157 rainfall stations (Cf. Fig. 18).

This study was carried out on a SIDA sub-contract
for Swedforest Consulting AB.

Address:

Swedforest Consulting AB
Box 154, S-182 12 Danderyd, Sweden
Tel. 08/7552755, Telex: 11758

AN ASSESSMENT OF WOODY BIOMASS, COMMUNITY FORESTS, LAND USE AND SOIL EROSION IN ETHIOPIA

A feasibility study on the use of remote sensing and GIS-analysis for planning purpose

CONTENTS **Page**

1. PREFACE 6.

1.1. Project aim 6.
1.2. Target group 7.
1.3. Team members 7.
1.4. Acknowledgements 8.

2. INTRODUCTION 9.

2.1. Background 9.
2.2. Summary of the project proposal 9.

2.2.1. Data needs 9.
2.2.2. Conclusions 10.
2.2.3. Project proposal 10.

2.3. Project start 12.

3. DATA AND ANALYSIS REQUIREMENTS FOR INTEGRATED PLANNING 13.

4. SOME ASPECTS ON DATA COLLECTION 13.

4.1. The traditional field based approach 13.
4.2. The remote sensing approach 14.
4.3. The remote sensing ability 15.
4.4. Additional data requirements for complex analysis 16.

5. SOME ASPECTS ON DATA ANALYSIS 16.

5.1. The need for computer support 16.
5.2. The Geographical Information System approach 17.

5.2.1. GIS description 17.
5.2.2. Data administration and generation of information 18.
5.2.3. Integrated analysis 19.
5.2.4. Natural resources supply-demand modelling 19.

6. METHODOLOGY AND DATA 20.

6.1. The three-level monitoring and planning approach 20.
6.2. Test areas 21.

6.2.1. Gojam focusing on Mertule Mariam 22.
6.2.2. Shewa focusing on Wilbareg 22.

6.3. Field data collection 23.
6.4. Collection of auxiliary data 24.
6.5. Land use/land cover assessment 24.
6.6. Woody biomass assessments 25.
6.7. Woody biomass supply/demand modelling 30.

6.7.1. Annual growth and demand assumptions 30.
6.7.2. Modelling on the national and regional levels 30.
6.7.3. Modelling on the local levels 34.

6.8. Soil erosion 34.

6.8.1. The Universal Soil Loss Equation 34.
6.8.2. USLE factor estimations 38.
6.8.3. Erosion modelling and assessment of conservation needs 42.
6.8.4. Severly denuded lands 43.

6.9. Result presentation 43.

7. RESULTS 49.

7.1. Introduction 49.
7.2. The Shewa case 49.

7.2.1. Land use 49.
7.2.2. Woody biomass supply/demand modelling 51.
7.2.3. Soil erosion modelling 51.

7.3. The Gojam case 59.

7.3.1. Land use 59.
7.3.2. Woody biomass supply/demand modelling 60.
7.3.3. Soil erosion modelling 61.

7.4. Integrated planning. A few examples. 67.

8. RESTRICTIONS 69.

9. COSTS 70.

10. REFERENCES 71.

Appendix 1. Terms of Reference 74.
Appendix 2. Satellite characteristics in summary 75.

AN ASSESSMENT OF WOODY BIOMASS, COMMUNITY FORESTS, LAND USE AND SOIL EROSION IN ETHIOPIA

A feasibility study on the use of remote sensing and GIS-analysis for planning purposes in developing countries

Dr. Ulf Helldén

1. PREFACE

1.1. Project aim

This is the final report of a feasibility study on the use of remote sensing and geographical information system analysis for community forest, land use and soil erosion assessments in Ethiopia. The study was carried out during 1986/87 on a 1-year contract for the Swedish International Development Agency (SIDA). The study covered approximately 40 000 km^2.

The project aim was to demonstrate the feasibility of a multipurpose monitoring and planning approach, for integrated natural resources planning, within the frames of a national strategic monitoring, information and planning system. The feasibility study was based on the integration of:

- a 3-level monitoring approach (space, air and ground monitoring levels) for multipurpose data collection
- a 3-level planning approach (national, regional and local/ management planning levels) for integrated natural resources planning.

The monitoring and planning system was expected to answer questions of the type listed below. The results should be in the form of continuously up-to-date maps and statistical tables. For a further discussion of this subject please refer to Section 7.4.

- Assuming it is a political goal that each community (Awraja) should rely on its own woody biomass resources, how many hectares must be afforested to reach a balance in supply-demand within 10 years, 15 years, 20 years?
- Considering the average crop yield and the land available for cultivation/household and for grazing (in relation to the minimum per capita area requirement), is there enough land to afforest (today, tomorrow)?
- How many seedlings and nurseries are needed/administrative unit?

What are the investments needed per region, per province and per community (Awraja) to reach the political goal?

-Is the woody biomass deficit Awraja suffering from untolerable high soil loss? If so, is it probable that soil conservation measures can increase the crop yield sufficiently in order to release land for afforestation?

-What are the annual investments needed (labour, tools, total costs) per Awraja or major drainage basin to reduce the soil loss to an acceptable level within 5 years, 10 years, 20 years?

The methodologies, models and results presented in the report should not be considered as the final technical result of a survey, but rather as a dynamic information technology approach, or idea, very much open to refinements and further development.

1.2. Target group

The report was primarily written for the following readers:

- Politicians, planners and decision makers in the developing countries with national reponsibilities related to the sustainable use of the renewable natural resources.

- Decision makers, and project planners at donor agencies and aid organisations with responsibilities related to the food, fuelwood and rangeland resources of the developing countries.

However, to give the technicians and environmental experts a fair chance to evaluate the report, technical discussions could not be completely avoided.

1.3. Team members

The team members responsible for the project were:

Ulf Helldén, Ph.D., Ass. Professor and Manager for the Lund University Remote Sensing Laboratory, Department of Physical Geography. Researcher and expert on Remote Sensing, Environmental Monitoring and Analysis with the emphasis on African Environments.

Ralph Jakobsson, Swedforest consultant, Project Management and expert on Forestry and Community Forestry.

Lennart Olsson, Ph.D., Researcher at the Lund University Remote Sensing Laboratory and expert on Remote Sensing, Environmental Monitoring and Analysis with the emphasis on African Environments.

The following persons at the Lund University Remote Sensing Laboratory contributed to the project:

Lars Eklundh, B.Sc., and Petter Pilesjö, B.Sc., carried out the stereo plotting for the generation of detailed topographical maps of Gojam. They contributed considerably to editing the digital elevation models and to generating the national precipitation data base.

Stefan Pinzke., B.Sc., programmer, contributed greatly to the software management and development.

1.4. Acknowledgements

The study could never have bene performed without the invaluable support of individual staff members of the following organisations in Ethiopia.

- State Forest Conservation & Development Department, NRCDMD
- Community Forest, Soil and Water Conservation Department, NRCDMD
- Land Use Planning Department, NRCDMD
- The Soil Conservation Research Programme, NRCDMD
- State Forest Inventory Section, NRCDMD
- DCO/SIDA Addis Abeba
- Ethiopian Mapping Agency
- National Meteorological Services Agency
- Central Statistics Office
- Wondo Genet Forestry Training Centre

To them and all those who provided assistance, the team wishes to extend its gratitude for all the support, kindness and hospitality received.

2. INTRODUCTION

2.1. Background

Sweden is supporting community forestry and soil conservation activities in various parts of Ethiopia. To study the possibility of strengthening the background information for planning purposes regarding these activities, as well as for land use and land use planning in a broader sense, SIDA allocated 2.9 million SEK (approx. 0.76 million Birr) in 1984 to initiate a feasibility study.

The feasibility study was divided into two phases;

- the first preparatory phase was to investigate information requirements within the present National Resources Conservation and Development Main Department (NRCDMD), to investigate possible land resources inventory and information analysis methods and to assess existing facilities in Ethiopia in terms of data available, manpower, equipment, training etc. and to suggest a work plan for phase II.

- the second operational phase was to study and demonstrate the feasibility and potential of the environmental monitoring and information analyses approach, suggested in the preparatory phase, focusing on two Ethiopian test areas in Shewa and Gojam respectively.

The Orgut-Swedforest Consortium was assigned to carry out the preparatory phase (Phase I) during one month in November/December in 1984 including approx. three weeks in Ethiopia. The assignment was reported at the end of December 1984 (Orgut-Swedforest 1984).

2.2. Summary of the project proposal

2.2.1. Data needs.

As far as community forests are concerned, there was no information available with the exception of scattered data on annual seedling production. The need for information about survival rates, present community forest distributions, woody biomass estimates, supply/ demand descriptions etc. was stressed by the NRCDMD officials.

Land use/land cover maps to the scale of 1:1 million were published recently by the Land Use Planning Department (FAO 1984a).The maps were based on manual interpretation of Landsat MSS imagery, recorded during 1972-1976, combined with field work. The time consumption was 6-7 years in order to generate a complete national coverage, implying that the information presented is a mixture of the land resources situation some 8-12 years ago.

Regarding soil erosion and soil conservation no reliable information was available concerning soil erosion rates, distribution of land severely affected by land degradation or the distribution and survival rate of implemented soil conservation measures according to the Soil Conservation Research Project (SCRP), Soil and Water conservation Department. The Community Forest, Soil and Water Conservation Department, NRCDMD, stressed the need for an integrated information system at a high administrative level to:

-compile, handle, analyse and dissiminate information on the land degradation and soil conservation situation at a national level.

-integrate erosion and soil conservation data with other types of information such as land use, crop yields, crop production, population distribution and pressure, grazing pressure and grazing resources, fuelwood supply-demand data etc., to assess real conservation needs and priorities.

The implementation of soil conservation/food supply programs is at present based on the strength of the crying voices rather than on an analysis and priority of the national needs.

2.2.2. Conclusions.

In summary, it was concluded that there is no reliable and up-to-date information, information retrieval system or information analysis system available in Ethiopia, on a national and regular basis, concerning either community forests, state forests, land use/land cover or soil erosion/soil conservation.

It was also concluded that the only feasible way to obtain and handle such information on a regular (annual) basis was to base it on the very latest developments in information technology and new research findings from the latest 3-4 year period regarding modern environmental monitoring and remote sensing technology applied to African environments.

2.2.3. Project proposal.

It was suggested that Phase II of the feasibility study should be based on a Geographical Information System (GIS) approach combining the latest findings in computer based remote sensing and data handling with visual photo (satellite and air photos) interpretation and field work. The aim was to study and demonstrate the potential of the approach, compared to existing and traditional options, for integrated environmental monitoring and planning.

The study should exemplify (simulate) a three level approach including the national, regional and local level to illustrate the potential of going from the generalized and annually updated national information level down to the detailed management level (Fig. 1). This "zooming" procedure should be carried out whenever a specific area is considered as a high priority inventory area.

3-LEVEL CONCEPT

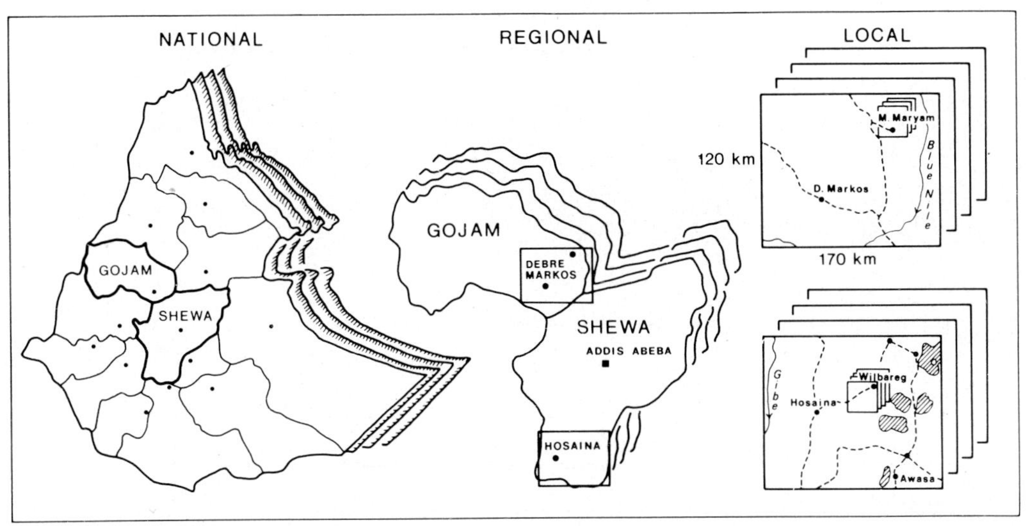

Fig. 1. The 3-level concept including the national, regional and local planning levels for monitoring from space, air and ground. The different information layers are stored and analysed in a Geographical Information System. The frames indicate the two areas studied in Gojam and Shewa respectively (40 000 km^2).

Inventory results to be expected should include distribution of soil erosion rates, severely degraded lands, land use/land cover, community forests and quantifications of standing fuelwood biomass.

Additional items to be demonstrated could include e.g. fuelwood supply/demand analysis, potential for trend analysis and simulations of the environmental impact of planning decisions.

The concept of the project approach is partly illustrated in Fig. 2.

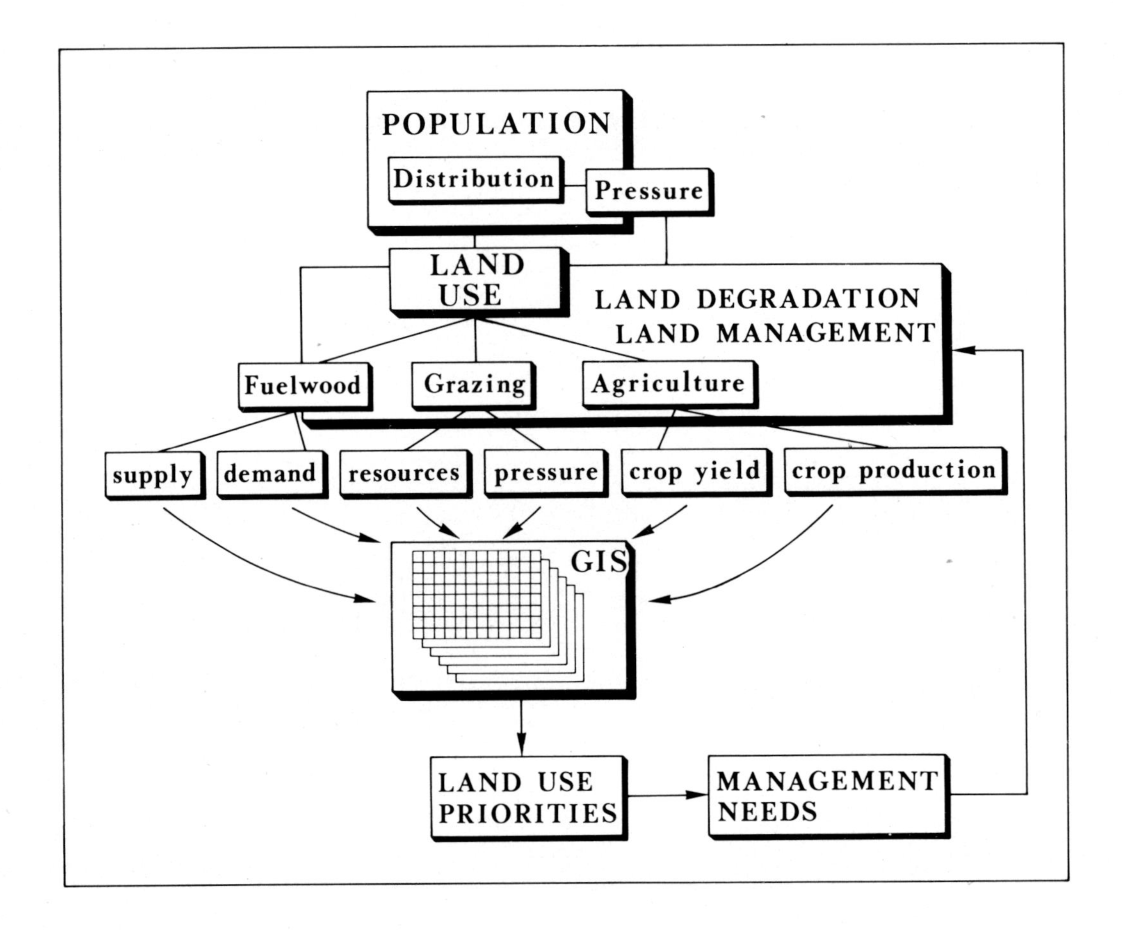

Fig. 2. An example of essential components that can be included in a natural resources strategic planning system. Woody biomass, community forests, land use and soil erosion are emphasized in this study.

2.3. Project start

In January 1986 SIDA assigned the Orgut-Swedforest Consortium to implement Phase II of the feasibility study. Dr. Ulf Helldén and Dr. Lennart Olsson, Remote Sensing Laboratory, Department of Physical Geography, University of Lund, Sweden were sub-contracted to carry out the study in accordance with the proposal. For a full description of the Terms of Reference please refer to Appendix 1.

Phase II was initiated at the beginning of February in 1986 when a Swedforest consultancy team arrived at Addis Abeba for a one month data collection and field work period. The field work, as well as the project plan, was summarized in a debriefing report distributed in March 1986 (Orgut-Swedforest 1986).

3. DATA AND ANALYSIS REQUIREMENTS FOR INTEGRATED PLANNING

Many people agree that the deforestation and desertification/ land degradation situation in Africa is serious, threatening existing production systems and possibly resulting in the generation of non-productive desert-like conditions and famines.

Knowledge about the actual magnitude, the speed, the trends, the spatial distribution, the environmental and socio-economic causes and consequences of deforestation and land degradation is however limited. The role played by climatic fluctuations and droughts in these processes is one of the main open questions of environmental impact. Knowledge regarding the present food, fodder (grazing resources) and fuelwood supply-demand pattern in individual countries is also limited, making realistic planning for the future a very difficult task.

It is obvious that the management of existing natural resources needs planning for a sustainable production (Cf. Fig.2). Investment in action programs, e.g. huge afforestation and land productivity rehabilitation programs, must be based on cost-benefit assessments in order to set up priorities. The priorities for action should not only be based on the expressed needs of a region (province, county, drainage basin or any optional area) but also on the expected benefit (environmental= sustained yield, political, socio-economic) compared to the assessed long term needs and benefits of other regions. This is an impossible task without national, easily accessible and up-to-date information about the natural resources supply-demand pattern, status, trends and spatial distribution. There is also a continuous need for information about the environmental and socio-economic impact of implemented action programs to learn from experience and to initiate corrections or alternative strategies when needed.

Information can only be obtained through the analysis of data including evidence from human experience. The quality of the information depends on the relevance of the analysis and on the data collection strategy.

4. SOME ASPECTS ON DATA COLLECTION

4.1. The traditional field based approach

A few common approaches to environmental data collection are mentioned below.

-Data and conclusions concerning natural resources related problems, are based on interdisciplinary "case studies", detailed but often with limited spatial coverage.

- Different sampling techniques when data and conclusion are based on the statistical theory of probabilities.

- Variations of "sampling techniques" lacking a sound statistical base. It is not uncommon that data and conclusions are based on non-systematic, subjective and biased observations by expert missions directed to travel through a pre-specified problem area over a short period of time.

The approaches mentioned can be referred to as point measurements from the national or regional point of view. They are detailed but with a questionable relevance for areas outside the points. They are much too slow and too work intensive to be applied at a national or even regional level on a monitoring basis.

Modern remote sensing offers an extremely powerful complement to the traditional tools for environmental data collection including a national and regional monitoring capability.

4.2. The remote sensing approach

During the past 15 years new and unique possibilities have been created to collect environmental data on a monitoring basis. Data can now be collected on an individual grid cell basis, in a grid net of optional size, with a complete ground coverage of any region of interest, be it a continent or a small development project area. Thousands of square kilometers can be covered in a matter of seconds, minutes or hours depending on the size of the individual cells in the grid net. It can be carried out by applying an adequate level of remote sensing technology, ranging from the use of air photography to the use of earth resources and meteorological satellites.

The African continent is covered every 30 minutes by the meteorological satellite Meteosat and it is covered on a daily basis by satellites in the NOAA series (grid cells 1-4 km). The minimum size of a grid cell measured by a commercial satellite today is 10 x 10 m in panchromatic mode and 20 x 20 m in multispectral mode (SPOT, launched February 1986).

Whenever environmental data are needed for retrospective analyses the use of historical remotely sensed data is often the only available solution. Useful satellite data (Landsat MSS) is available from 1972 onwards.

Please refer to Appendix 2 for a summary description of the satellites mentioned.

4.3. The remote sensing ability

Useful data can be acquired, on a monitoring basis, through satellite based remote sensing technology within the following major fields of interest for the documentation and analysis of natural resources and for planning purposes in African environments. It should be noted that field work, usually carried out on a sample approach based on the remotely sensed data, is needed in most cases to transform the data into valid information.

- Creation of statistical area sampling frames, based on a combined land use/land cover stratification, to be used in e.g. a national agricultural census to increase the accuracy of estimates and decrease the costs.

- Geology and geomorphology:
  soils, bedrock, landforms, geological structure, soil surface moisture to a certain extent

- Hydrology:
  drainage patterns, surface water and water turbidity

- Land use and cover:
  rain-fed and irrigated cultivations, fallow, clearings, burnt areas, grazing land, villages, towns

- Population:
  density and distribution to a certain extent

- Infrastucture:
  roads, tracks

- Vegetation cover and types:
  desert-grassland-savanna woodland-forests, dominating species composition, field layer coverage, canopy coverage

- Physical indicators of soil erosion and desertification:
  desert margins, desert patches, dust plumes, gullies, badlands, flood deposits, alluvial fans, slides, barren-degraded land, dunes

- Landscape green biomass situation:
  green biomass status, seasonal and annual green biomass accumulation and productivity, possibly crop yields

- Fuelwood resources:
  standing woody biomass in tons/ha for specific environments

- Climate related factors:
  albedo (i.e the percentage of the insolation incident upon a surface which is reflected back towards space. It is of major

importance to study the long-term impact of deforestation and desertification on climate), temperature, possibly patterns of precipitation and evaporation.

The transformation of satellite data into valid information is often carried out through a combination of visual (manual) data interpretation and computer processing. However, visual data interpretation alone is often satisfactory. Computer processing, on the other hand, is usually needed when transforming the original data into quantitative information (e.g. standing woody biomass or green biomass productivity expressed in tons/ha).

4.4. Additional data requirements for complex analysis

Much of the environmental data requirement for planning can be satisfied on a monitoring basis by employing an adequate remote sensing technology (satellite data/air photos). However, the need for integrated planning calls for additional and supplementary data. Some examples of important data that might be available or have to be collected through traditional field work are statistics on livestock composition, distribution and feeding requirements, data on soil fertility, water availability and tree species productivity. Of equal or greater importance are statistics on population distribution, agricultural production and productivity and detailed statistics on climate with special reference to precipitation characteristics. Information on the local variations of man's perception of his environment, his strategy and options for survival and his interaction with the socio-economic and political environment creates a need for even more data to be collected by conventional means.

How to handle all these data, if and when available? How to merge the remotely sensed environmental data with data or information from other sources for the integrated analysis required to manage a sustained land productivity and plan for the future?

5. SOME ASPECTS ON DATA ANALYSIS

5.1. The need for computer support

Several documentation problems can be solved through visual interpretation of satellite imagery, e.g. the assessment of the distribution and growth of desert patches and severely degraded lands or the assessment of areas covered by forest and the speed of deforestation. However, the more complex transformation of satellite data into adequate information and the need for system's analysis of the complex "man-environment" interaction, involving huge amounts of data from different sources, calls for computer support.

The fact that the environmental data recorded by the earth resources satellites are stored and available in digital and computer compatible form from the very beginning enhances the possibility of allowing computers to complement man for an optimal use of the data.

5.2. The Geographical Information System Approach

5.2.1. GIS description.

Environmental information presented in map form is a necessary tool for the planning and management of natural resources, as well as for research on the distribution and allocation of resources. Maps can be seen as a means for communication between researchers, decision makers and planners. The amount of information that can be presented in map form is tremendous. Both status, trends and projections can be presented in a conceptually simple way. Cartography has been one of the cornerstones of geography since the very beginning of the discipline. To keep pace with the increasing capacity to collect environmental data through remote sensing and the increasing demands of supplying data to users of all categories, the conventional data handling methods must be supplemented by modern computer assisted techniques.

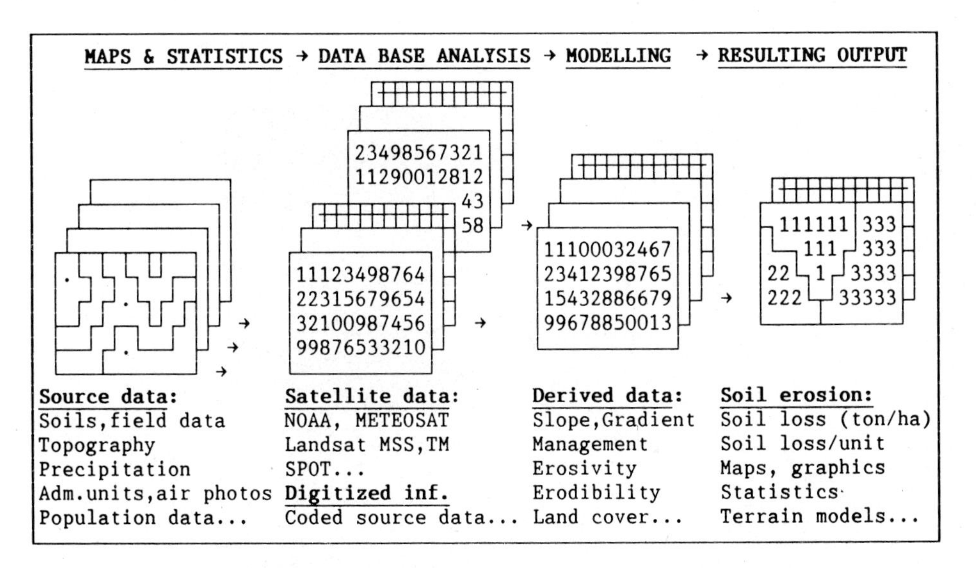

Fig. 3. The principle of a geographical information system (GIS). Spatial information (e.g. thematic maps and remotely sensed data) are stored as different information layers in a grid system. Tabular data can be linked to the spatial information layers.

Developments in geography applying recent achievements in computer science, remote sensing, image processing and computer cartography have created an instrument, the geographical information system (GIS), with an unique potential to optimize environmental system's analysis and documentation, when combining remotely sensed data with information from other data sources.

A Geographical Information System (GIS) is a multipurpose computer based information system for retrieval, administration, processing, integrated analysis and graphic/cartographic and statistical presentation of any kind and combination of data/information which can be defined in time and space. The principle is outlined in Fig. 3. As an example, the figure illustrates how a soil erosion model can be derived.

5.2.2. Data administration and generation of information.

Possible types of input data include remotely sensed data in digital form, e.g. satellite data, to be used for further image processing and classification purposes for the generation of new information layers such as land use/land cover distributions, green or woody biomass distributions etc. The satellite data, or the generated information layers, can be combined with digitized information e.g. topographical data for the creation of digital terrain models, population data, socio-economic data, crop yields, livestock distribution for the assessment of grazing pressure, soil fertility data and so on.

Fig. 4. An image processing and GIS workstation at the Lund University Remote Sensing Laboratory.

Information from spatially limited case studies and data from scattered point measurements, e.g. precipitation data from a national net of climatic stations, can be interpolated and linked to existing data bases in optional scales.

All operations are carried out in an interactive environment through a continuous dialogue between the operator and the computer system (Fig. 4). Maps, statistics, diagrams etc. can be immediately displayed and edited before the final printing.

5.2.3. Integrated analysis.

The major advantage with a GIS is the possibility to integrate and analyse very large amounts of data from different sources and with different themes for computer based generation of new information layers, maps and statistics for planning purposes. The information available can be presented in optional combinations.

Based on political priorities, the planner can define the rules e.g. for the identification of the areas (regions, provinces, communities, drainage basins) with the highest needs for soil conservation for an estimation of the cost/benefit relationship in each of the areas selected. The characteristics selected to identify the areas might be any combination of a high population density interval, a low available land/capita ratio interval, a high, but declining, annual crop yield, a low per capita income, a high rain storm frequency, a medium soil loss interval, a woody biomass deficit and a specified slope interval.

It would be a tremendous and practically impossible task to carry out such operations manually by combining and comparing map sheets, with different themes, scales and ages, with each other.

Another GIS advantage is the possibility to continuously up-date the information in the system on a monitoring basis, e.g. through satellite data analysis, for change studies, trend analysis and the generation of prognosis.

The potential for dynamic simulation and modelling is a third advantage described below.

5.2.4. Natural resources supply-demand modelling.

Remote sensing methods are frequently used for monitoring the supply of natural resources, e.g. agricultural yield, grazing and wood resources. However, the ultimate goal of resource monitoring must be to go one step further; **to analyse not only the supply but also the demand and the accessibility of the resources**.

Development of spatial models of land use may be a way towards the ultimate monitoring system, where supply, demand and accessibility of resources can be analysed in an integrated fashion. Combined models describing supply and demand have been applied to studies of wood resources and millet production at the Lund University Remote Sensing Laboratory (Olsson,L 1985, Olsson,K 1985b).

In the case of wood resources, the modelling was based on remotely sensed data describing woody biomass supply (tons/ha) and estimated annual consumption of wood(interviews), in Kordofan province of the Sudan. The modelling compared the demand for each village with the resources within a certain walking distance. As a result the walking distance needed to satisfy the demand or the actual wood deficit was calculated for each village. The methodology is readily applicable to studies on future development of the wood situation, through simulation of different variables, population increase, migration, afforestation, land management etc.

In the case of crop production, the modelling was based on population distribution and statistical prediction of millet productivity in the Sudan. Spatial modelling yielded detailed information on land availability (Ha/ capita). The combination of productivity estimates and land availability made it possible to project the per capita production. When combined with more accurate measurements of actual crop return, this approach will be a powerful instrument for studies of spatial distribution of per capita food production.

The multipurpose inventory/monitoring and modelling techniques described below, with special reference to community forest, land use and soil erosion assessments, are designed to fit into the GIS approach described.

6. METHODOLOGY AND DATA

6.1. The three-level monitoring and planning approach

From the practical and economical point of view it is an almost impossible task to acquire detailed <u>and</u> up-to-date information about the national resources of a large country like Ethiopia. If "detailed" information was available it would be an equally almost impossible task to keep it continuously updated.

The present production of topographical maps in Ethiopia illustrates the problem. It takes more than 1 700 map sheets, of the scale 1:50 000, to cover Ethiopia (1.2 million km^2). The annual production at the Ethiopian Mapping Agency, employing 350 people, was about 30 topographical sheets (1:50 000) in 1985 (Andersson and Ekelund, 1985). With this speed more than 50 years (17 500 man-years) are needed to acquire a complete national coverage.

It is obvious that detailed and up-to-date information, needed for natural resources planning and management, can only be obtained for very limited high priority areas. Regional/province strategies and priorities must be based on less detailed and less accurate data collection approaches to cover much larger areas with information of immediate interest. This in turn creates the need for a somewhat more generalized, but always up-to-date, information level to form the basis for national strategies and priorities. The three different information and planning levels (corresponding to different data collection levels) are referred to as the local, regional and national levels in this study.

The principle of the data collection and information concept is to go step by step from the generalized (national) level to the detailed (local) level directing the investments in data collection and analysis to priorities based on the information available at the preceeding level (Cf. Fig. 1).

The following scales and remotely sensed data (always combined with field data) illustrate the concept of the three-level monitoring and planning approach:

-The national level, 1:500 000 - 1:250 000 employing Landsat MSS data on a multipurpose monitoring basis (once/1-5 years).

The information generalization level 1:250 000 was used in this study.

-The regional level, 1:250 000 - 1:100 000, employing Landsat MSS data on a multipurpose monitoring basis (once/2-5 years) or priority basis (possibly employing Landsat TM data).

The information generalization level 1:250 000 and Landsat MSS data were used in this study.

-The local level, 1:50 000 - 1:10 000, employing Landsat TM data, SPOT data, air photos and field work depending on the information needed. The decision to go into the local level (a one time single theme or multipurpose survey or monitoring during a defined period) is always based on priorities set up in the preceeding level.

The information generalization levels 1:50 000 (Landsat TM and SPOT data), 1:20 000 (air photos) and 1:10 000 (air photos) were used in this study.

6.2. Test areas

Two Ethiopian test areas were selected to simulate the three-level monitoring and planning approach focusing on land use, woody biomass and soil erosion (Cf. Fig. 1). Each area corresponds to the coverage of a 1:250 000 topographical sheet (approx. 167 x 115 km, corresponding to 24 1:50 000 sheets) illustrating the generalized regional (and possibly national) level.

One 1:50 000 sheet (approx. 27.5 x 27.5 km) was selected within each area to illustrate the local level.

Parts of the Gojam test area were finally covered by 1:20 000 (approx. 8 x 20 km) and 1:10 000 (approx. 6 x 6 km) information layers to illustrate alternative local/management levels.

6.2.1. Gojam focusing on Mertule Mariam.

Topographical Maps: (Ethiopian Mapping Agency)
1:250 000 sheet, EMA3, sheet NC 37-6, Debre Mark'os
1: 50 000 sheet, 1038 A2, not printed, two new sheets were generated by the project for the original scales of 1:10 000 and 1:20 000 (contour intervals at 10 m and 25 m respectively).

Landsat data: (EOSAT, USA)
MSS and TM, Path/row 169/52 & 169/53, Febr. 26, 1985

Air photos: (Ethiopian Mapping Agency)

1:40 000,	Jan. 1982	1:20 000., Febr. 1985
ET 2:2 S3	10:0676-0685	R1 S01 007-018
2:2 S3	11:0708-0716	R1 S02 019-035
2:2 S3	12:0739-0747	R1 S03 036-052
2:2 S3	13:0770-0779	
2 F05	14:727-736	
2 F05	15:831-839	

The area studied is under intensive agriculture, characterized by highland plateaus dissected by streams and river gorges. Soil erosion is severe and some terracing efforts have been made.The area is fairly poor in woody biomass resources. The resource available, beside the Acacia dominated canyons, consists of community forests and peasants trees (Eucalyptus spp.). The SIDA supported Community Forestry Centre is under establishment within the Mota Awraja using Mertule Mariam as a project centre.

6.2.2. Shewa focusing on Wilbareg.

Topographical maps:
1:250 000 sheet: EMA3, sheet NB 37-2, Hos'aina.
1: 50 000 sheet: 0738 A3, Wilbareg.

Landsat data:
MSS, Path/Row 168/55 & 169/55, Dec. 18, 1984
MSS, Path/Row 169/55, March 14, 1985
TM, Path/Row 168/55:1, Dec.01, 1984

TM, Path/Row 169/55:2, March 14, 1985

SPOT data: (Satellite Image, Kiruna, Sweden)
Multispectral mode, Path/Row 137/335, Jan.20,1987

With the exception of the Rift Valley bottom, this is considered to be one of the most densely populated parts of Ethiopia, and is under intensive agriculture. Soil erosion problems are increasingly felt and some areas are already severely degraded with vast badlands. Soil conservation measures and planting activities have been undertaken and the community forest coverage is very high in parts of the area.

6.3. Field data collection

Field data is needed to calibrate remotely sensed data to the environment under study. During February 1986 the team spent almost 20 days collecting field data through a sampling approach based on coding some 13 environmental parameters once per kilometer from two vehicles transversing the areas of interest on most roads and tracks available. The parameters included vegetation type, land use, canopy cover, tree height, field layer coverage, dominant tree/bush species, grazing intensity, erosion indicators, soil type and colour. Approximately 1800 observations were sampled and plotted on topographical maps along 1800 km of roads and tracks within the Shewa and Gojam test areas. In total the team travelled some 3500 km in Gojam and Shewa.

Acacias, ranging from 100-1500 kg wet weight were measured (crown diameter and height), cut into pieces and weighed to be merged with an already existing data set (Fig. 5). The set was used for satellite data calibrations (please refer to section 6.6).

Fig. 5. Destructive measurements (crown diameter, height and weight) of Acacias in Ethiopia for satellite data calibrations.

Eucalyptus trees ranging up to 20 m height were measured, cut and weighed. Seven 100 x 100 m areas with a varying canopy cover were identified and field measured regarding tree/bush height and crown diameter for woody biomass estimates and later calibration to air photos and satellite data.

6.4. Collection of auxiliary data

Data collected in Addis Abeba included:

- all monthly precipitation data tabulated for 157 stations all over the country

- all precipitation intensity data available to be used for the soil erosion assessments.

- topographical maps and air photos for the generation of digital terrain models as one important input for the soil erosion assessments.

- important data concerning soil erodability factors and rainfall erosivity from the Soil Conservation Research Project attached to the Community Forest, Soil and Water Conservation Department.

- available national statistics on crop yields, area production, the results of the 1984 population census as well as data on livestock distribution from the Central Statistics Office.

- most of the maps (national coverage) and technical reports produced by the FAO assistance to the land use planning project at the Land Use Planning Department.

Time limitations and Terms of Reference excluded food and grazing resources supply/demand modelling from the study. The available data on crop yields, area production and livestock were not used.

6.5. Land use/land cover assessment

Computer based classifications of Landsat MSS data, combined with visual interpretation and manual class delineation in direct interaction with the information system, were employed for the land use and land cover assessments at the scale of 1:250 000. Two digital mosaics, each one composed of two Landsat scenes, were generated to cover the two 1:250 000 sheets (one Landsat scene covers 185 x 185 km). All data were geometrically registered to the UTM coordinate system. The Landsat MSS data, recording the ground in each 56 x 79 m cell, was resampled to 100 m (1 ha) cells (picture elements, pixels). The resampling procedure can be considered as a generalization of the potential information content in the data set, speeding up the data processing through a considerable reduction of the data amounts.

A corresponding approach was used for the 1:50 000 coverage employing Landsat TM data (pixel size 30 x 30 m) resampled to 25 m pixels.

Air photo interpretation and stereo plotting (Wild Stereo Aviograph B8 S) were employed to generate the detailed topographical and thematic maps in the Mertule Mariam area in Gojam (original scales 1:10 000 and 1:20 000, contours at 10 m and 25 m intervals respectively).

Please refer to the results and map legends for a presentation of the land use and land cover classes used.

6.6. Woody biomass assessments

It has been demonstrated in several studies that it is possible to assess total green biomass (field layer and canopy), canopy cover and woody biomass through processing multispectral digital satellite data (Helldén and Olsson,K 1982, Olsson,K 1985b, Tucker et al. 1985, Justice 1986).

The green biomass assessment is based on the fact that chlorophyll absorbs red light for the photosynthesis and reflects energy in the near-infrared part of the spectrum in proportion to the amount of chlorophyll (greenness) available. However, the size of the absorbtion and reflection of a tree/bush canopy is also dependent on other canopy characteristics. Factors of very great importance are proportions of stems, branches and shadows per unit area.

If satellite data are recorded during the dry season (Jan-Feb.), when nothing but the tree/bush canopy is still green, the average canopy absorption and reflection properties (including leaves, branches, stems and shadows) are measured by the satellite in 30 x 30 m squares on the ground (Landsat TM data). A 185 x 185 km area is measured within a 25 seconds period. The data set covering such an area is made up of some 38 million 30 x 30 m squares each one containing the information about the absorption/reflection properties of the crown cover in that specific square (pixel).

The following procedure was used to establish a relationship between satellite data and the canopy cover and standing woody biomass respectively in every pixel.

-1. The data resulting from the destructive measurements of the Acacias in Ethiopia was merged into a corresponding set of bush and tree data collected in the Sudan. Relevant information about the Sudan study was presented by Helldén & Olsson,K 1982, and Olsson,K 1985a, 1985b.

It was found that the wet weight of an Acacia, ranging in size from a few kilograms up to almost 1.5 tons, can be described as a function of the crown diameter (Table I and Fig. 6a). The logarithmic relationship indicated in Fig.6b. was used in the study.

Since the crown diameter of individual trees and bushes is highly correlated to the weight of each individual, it is most probable that the total weight of any area can be described as a function of its canopy cover up to a certain level. That this assumption is correct was clearly demonstrated in the case of the Sudan (Olsson,K 1985a, 1985b). The assumption was verified as described in the next step.

-2. The total canopy and wet weight of seven 1 ha areas in the grasslands and woodlands of the Ethiopian Rift Valley was assessed on the ground. The resulting relationship between wet weight, expressed in ton/ha, and canopy cover, is illustrated in Fig. 7.

-3. The results presented imply that the standing woody biomass of any area can be assessed as soon as the canopy cover is known (for Acacia dominated grass-, bush- and woodlands). A dry season relationship between Landsat MSS data and canopy cover is illustrated in Fig. 8. It is based on the 7 field measured plots in Ethiopia merged with 29 plots in the Sudan. The Sudan canopy cover data were assessed in high resolution air photos (Olsson 1985a, 1985b).

Table I. Correlation matrix based on destructive measurements of 41 trees and bushes in the Sudan (32) and Ethiopia (9). The data is made up of a mixture of Acacia albida, A. mellifera, A. senegal, A. seyal, A. tortilis, Albizza amara, Balanites aegyptica represented in approx. equal proportions.

	HEIGHT	D	WEIGHT	D^2	LOG W
DIAMETER (D)	.897				
WEIGHT (W)	.818	.899			
D^2	.853	.952	.982		
LOG W	.866	.905	.691	.759	
LOG D^2	.856	.926	.691	.777	.971

D=crown diameter, W=wet weight

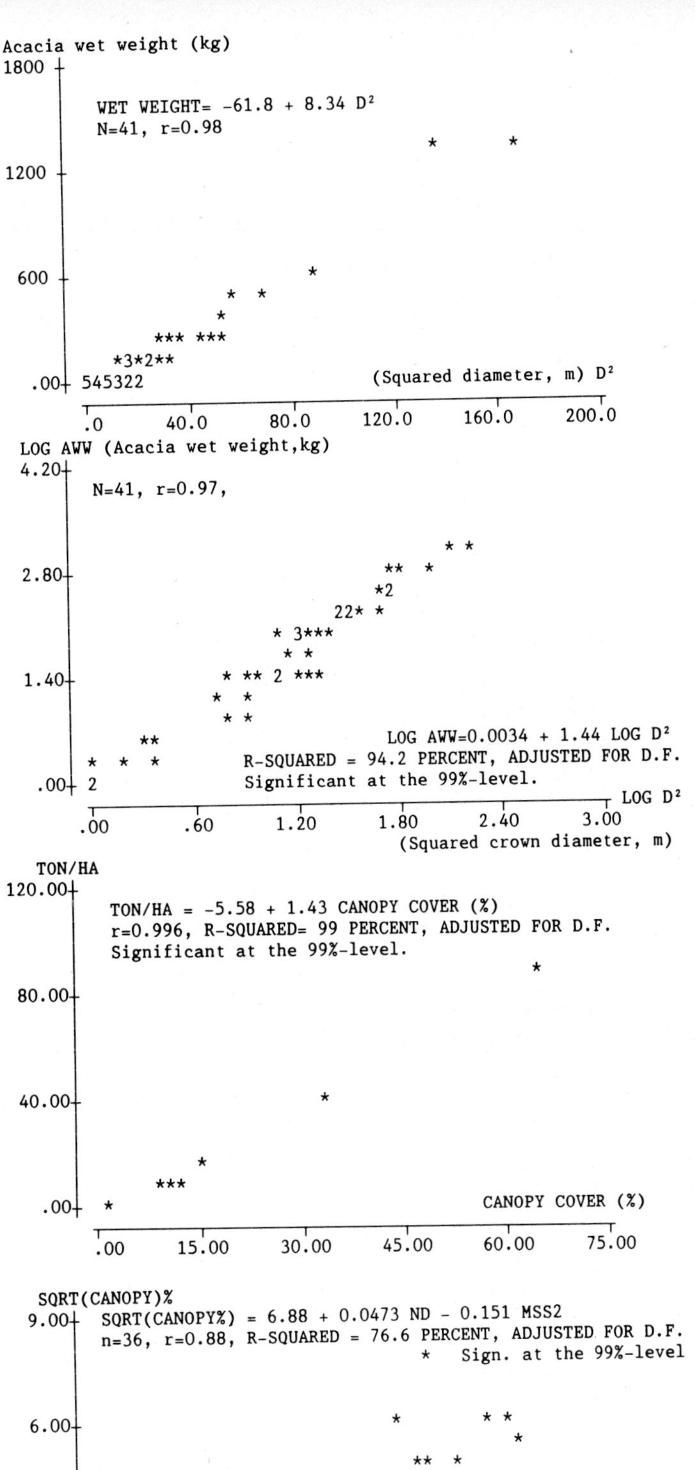

Fig. 6a. The relationship between wet weight(kg) and squared crown diameter (m) according to destructive measurements in the Sudan (32) and Ethiopia (9).

Fig. 6b. The log relationship between wet weight and squared crown diameter according to destructive measurements in the Sudan (32) and Ethiopia (9).

Fig. 7. The relationship between wet weight (kg) and canopy based on field measurements in seven 1 ha test areas in Ethiopian grasslands and woodlands.

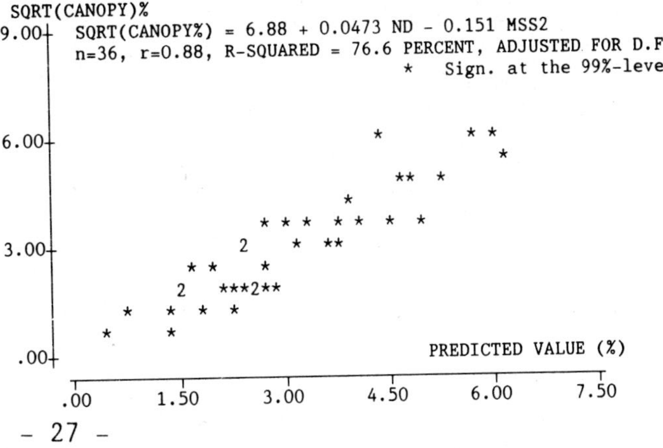

Fig. 8. Canopy cover as a function of Landsat MSS data (ND vegetation index and MSS 2) based on 36 calibration plots in the Sudan (29) and Ethiopia (7).

A corresponding dry season relationship between Landsat TM data and field measured canopy cover is presented in Fig. 9. It includes the Ethiopian data only. The satellite data were represented in both cases by the normalized difference vegetation index (ND) (the difference of the red and near infrared bands over the sum of the same bands) and the red MSS (MSS 2) and TM (TM 3) bands respectively. The ND was included to measure the "greenness" and the red bands to measure the non-green characteristics (shadows, stems, branches) of the canopy.

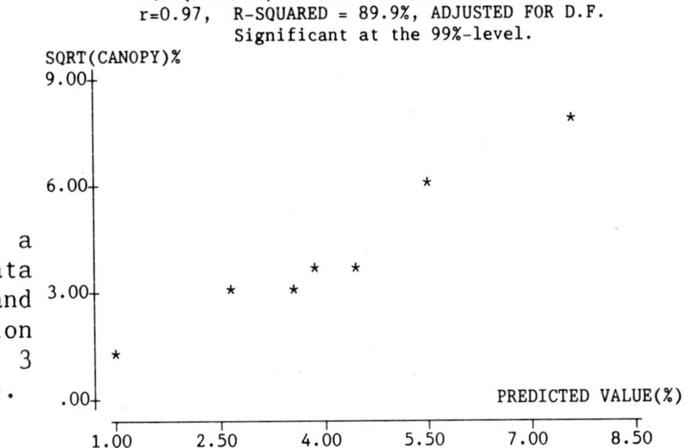

Fig. 9. Canopy cover as a function of Landsat TM data (ND vegetation index and TM3) based on 7 calibration plots in Ethiopia (step 3 in the working procedure).

It should be noted that no radiometric corrections were applied to any of the satellite data sets used. It implies that the regression models presented are only valid for the present data sets and environments. The models have to be modified in a possible operational phase employing relative radiometric calibration and corrections of the satellite data for differences in atmosphere, sun angle and solar irradiance. Methods to transform satellite digital grey levels into spectral radiance and at-satellite reflectance were summarized by e.g. Ahlcrona (1987), Hall-Könyves (1987), Markham and Barker (1986).

The woody biomass assessment and monitoring approach described above is valid for Acacia dominated tree grasslands and woodlands only. When it comes to forests (closed canopy), satellite data can only be used for delineation, stratification/dominant species classification and area change studies. Quantifications must be based on air photo (stereo models) interpretation, photogrammetry (tree height measurements) and field work (sampling basis or full coverage depending on size and priorities).

The overall, dominant, scattered forest-stand resource of Ethiopia, beside the Acacia spp. is made up of Eucalyptus spp. The Eucalyptus usually grows in very dense and small stands. These stands were identified and delineated and the areal distribution assessed by means of Maximum-Likelihood classifications of satellite data. Knowing, or assuming, the average number of stems/area and assessing the average height (in the field, from age or from air photos) of the stands, the standing woody biomass resource was assessed by employing the weight-height model illustrated in Fig. 10b. To simplify the data processing the average Eucalyptus height 10 m and the average Eucalyptus densities 1.5 stems/m^2 and 0.7 stems/m^2 were used for the Shewa and Gojam cases respectively.

However, a general Ethiopian plantation recommendation stipulating an Eucalyptus density corresponding to 0.44 stems/m^2 was found at a later stage.

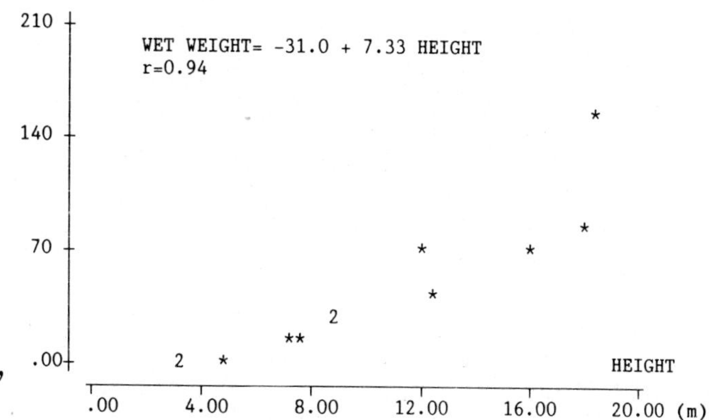

Fig. 10a. The relationship between height (m) and wet weight for 12 Eucalyptus Globulus cut in Wondo Genet, Ethiopia (Febr. 1986).

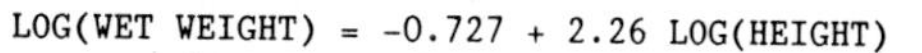

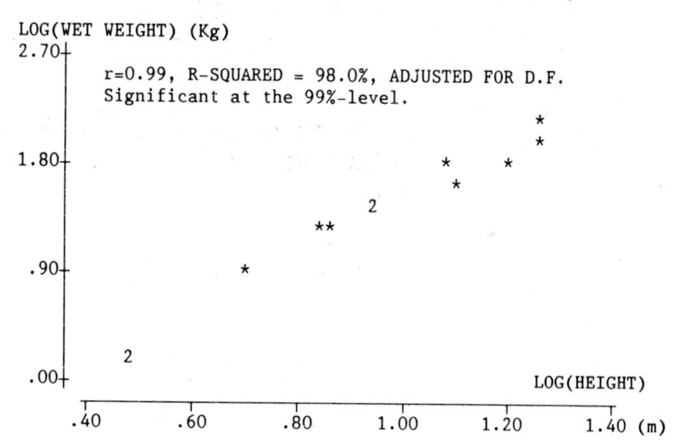

Fig. 10b. The log relationship between height and wet weight for 12 Eucalyptus Globulus cut in Wondo Genet, Ethiopia (Febr.-86).

6.7. Woody biomass supply/demand modelling.

6.7.1. Annual growth and demand assumptions.

Supply/demand modelling for planning purposes requires information about the size and the spatial distribution of the resources. It also requires information about the annual growth of the resources, assuming that nothing but the growth should be consumed for a sustained production.

The annual woody biomass growth (productivity) was assumed to be 10% of the standing biomass for Acacia and 20% for Eucalyptus. These figures can however easily be replaced by any suitable productivity figures, taking soils and climatic factors into account, for dynamic and interactive supply modelling.

Fuelwood consumption is considered to be the major consumption of wood in most African countries. The annual per capita consumption in a number of countries is presented in Table II. The non Ethiopian data is based on interviews and a literature review by Olsson,K (1985b). The urban consumption (including large villages) is considered to be in the range of 0.7-2.9 tons for the non Ethiopian countries presented in Table II (Olsson,K 1985 b).

Table II. Annual rural per capita consumption (tons) of fuelwood in some African countries.

Country	Consumption (ton/capita,year)	Source
Sudan	1.0 (0.5-1.4)	8 different sources, Olsson,K (1985b)
Chad	0.8	Howe and Gulick (1980)
Senegal	0.5	- " -
Niger	0.5	- " -
Mali	2.1	- " -
Ethiopia	0.7 m^3 (≃0.5 ton)	Kir (1985)

The annual per capita consumptions 0.6 and 1.0 tons were used as minimum and maximum values in the supply-demand modelling described below.

6.7.2. Modelling on the national and regional levels.

Several approaches were used for the demand modelling depending on the quality (resolution) of the population distribution input data and on the level of generalization wanted (dependent on the cost/ benefit relationship).

Tabulated population data from the 1984 national census was used as input data for the simulation of the national and regional levels (the 1:250 000 sheets) (Central Statistics Office 1985). Administrative boundaries were digitized and the population totals for Awrajas (provinces) and Weredas (counties) were divided by the true areas and the habitable areas (excluding lakes, steep canyon sides, marshes etc.) respectively yielding the number of people/pixel or the number of people/any optional land unit (Fig. 11-12). The difference between the supply (differentiated on pixel level) and the demand (differentiated on Wereda level) was used to illustrate the spatial distribution of surplus and deficit Weredas. The difference was also used to indicate the possible distribution of surplus and deficit areas within Weredas (Fig. 13).

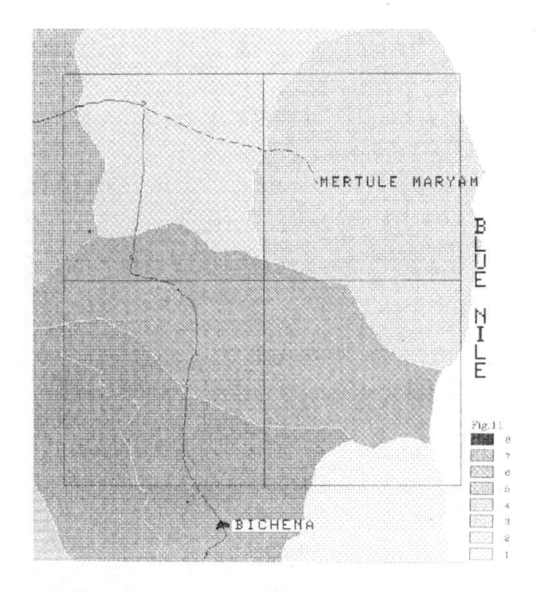

Fig.11. The average population density (inhabitants/ ha) of some of the counties (Weredas) bordering the Blue Nile river in Gojam. Each frame indicates the size of a 1:50 000 topographical map (approximately 27.5 km x 27.5 km). The shades correspond to:

1: 0.73 inh/ha, 73 inh/km²
2: 0.82 82
3: 0.84 84
4: 0.92 92
5: 1.04 104
6: 1.13 113
7: 1.23 123

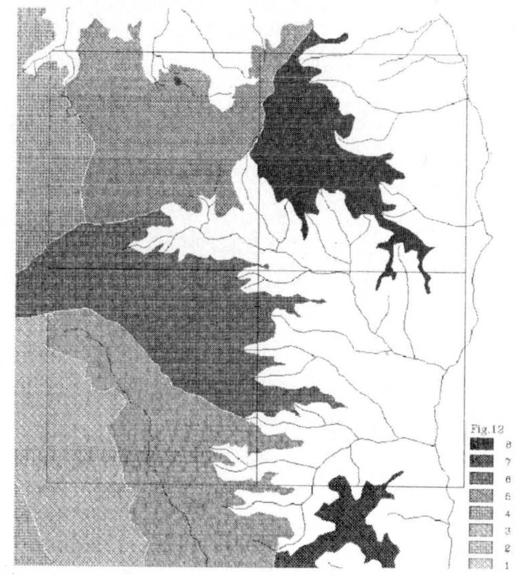

Fig.12. The population density of the habitable areas of the same counties, excluding the very steep canyons of the Blue Nile. Please notice that the population density increased up to 4 times for individual Weredas. The shades correspond to:

1: 1.20 inh/ha, 120 inh/km²
2: 1.40 140
3: 1.50 150
4: 1.70 170
5: 1.80 180
6: 2.00 200
7: 3.10 310
8: 3.50 350

Table IIIa. Example of a printout from the Shewa case data set listing county (Wereda) areas, population density, average standing woody biomass (according to two different satellite based models) and two assumed woody biomass annual productivities (10% and 20% respectively of the standing woody biomass). The first estimate (1) is based on the canopy cover - Landsat ND vegation index and MSS 2 illustrated in Fig. 8. The second estimate (2) is based on a corresponding canopy cover - Landsat MSS 2 relationship, yielding the same correlation coefficient (r=0.9)

WEREDAS (KENBATA & HADYA)	AREA(HA)	INH/HA	TON/HA(1)	TON/HA(2)	10%(1)	20%(1)
1. LIMU	90844	2.58	28.8	23.7	2.88	5.76
2. ANGACHA	40525	3.04	40.3	33.7	4.03	8.06
3. KEDIDA & GAMELA	30238	4.19	29.2	25.3	2.92	5.84
4. SIKE	51344	2.83	29.9	27.0	2.99	5.98
5. KACHA BIRA	26144	3.93	52.2	44.8	5.22	10.44
6.* OMO SHELEKO	79991	1.34	60.5	51.6	6.05	12.10
7.* TIMBARO	90836	2.08	41.5	35.3	4.15	8.30
8. KONTEB	20275	1.98	45.6	39.2	4.56	9.12
WEREDAS (HAIKOTCH & BUTAJIRA)						
9.* ZUWAI & AKABABI	94778	0.65	11.6	13.8	1.16	2.32
10. ARSSI & NEGELE	89050	1.05	45.6	37.9	4.56	9.12
11. SHASHEMENE	75081	2.00	57.4	50.2	5.74	11.48
12. SIRARO	124838	1.00	36.0	34.0	3.60	7.20
13. ALABA	121925	1.10	24.0	21.8	2.40	4.80
14. DALOCHA	61225	1.68	23.6	20.7	2.36	4.72
15. LANEFERO	87119	0.87	15.1	13.6	1.51	3.02
16.* SILITE	50768	1.84	25.4	22.3	2.54	5.08
17.* MESKAN & MAREKO	79805	2.20	18.6	15.7	1.86	3.72

*= Parts of these Weredas fall outside the 1:250 000 sheet

Table IIIb. Example of a printout from the Shewa case data set listing the difference between supply and demand (per ha and Wereda totals) for the two models mentioned in Table IIIa. The annual consumption was assumed to be 1 ton/capita. The woody biomass resource available for consumption was assumed to be 10% of the standing woody biomass.

WEREDAS	SUPPLY-DEMAND TON/HA (1)	SUPPLY-DEMAND TON/HA (2)	SUPPLY-DEMAND TON (1)	SUPPLY-DEMAND TON (2)
1. LIMU	0.30	-0.21	27253	-19077
2. ANGACHA	0.99	0.33	40120	13373
3. KEDIDA & GAMELA	-1.27	-1.66	-38402	-50195
4. SIKE	0.16	-0.13	8215	-6675
5. KACHA BIRA	1.29	0.55	33726	14379
6. * OMO SHELEKO	4.71	3.82	376758	305566
7. * TIMBARO	2.07	1.45	188031	131712
8. KONTEB	2.58	1.94	310310	233334
9. * ZUWAI & AKABABI	0.51	0.73	48337	69188
10. ARSSI & NEGELE	3.51	2.74	312565	243997
11. SHASHEMENE	3.74	3.02	280803	226745
12. SIRARO	2.60	2.40	324579	299611
13. ALABA	1.30	1.08	158503	131679
14. DALOCHA	0.68	0.39	41633	23878
15. LANEFERO	0.64	0.49	55756	42688
16. * SILITE	0.70	0.39	35538	19800
17. * MESKAN & MAREKO	-0.34	-0.63	-27134	-50277

*= Parts of these Weredas fall outside the 1:250 000 sheet

The total magnitude of the surplus and deficit in each Wereda was assessed and tabulated for varying productivities and per capita consumptions (Table IIIa-b). The results might as well have been presented in terms of overpopulation and underpopulation for any administrative unit of interest and expressed as total surplus or deficit of people. Or it could have been presented as the number of plants and nurseries needed to reach a balance in supply-demand in each administrative unit, including the costs, under different assumptions.

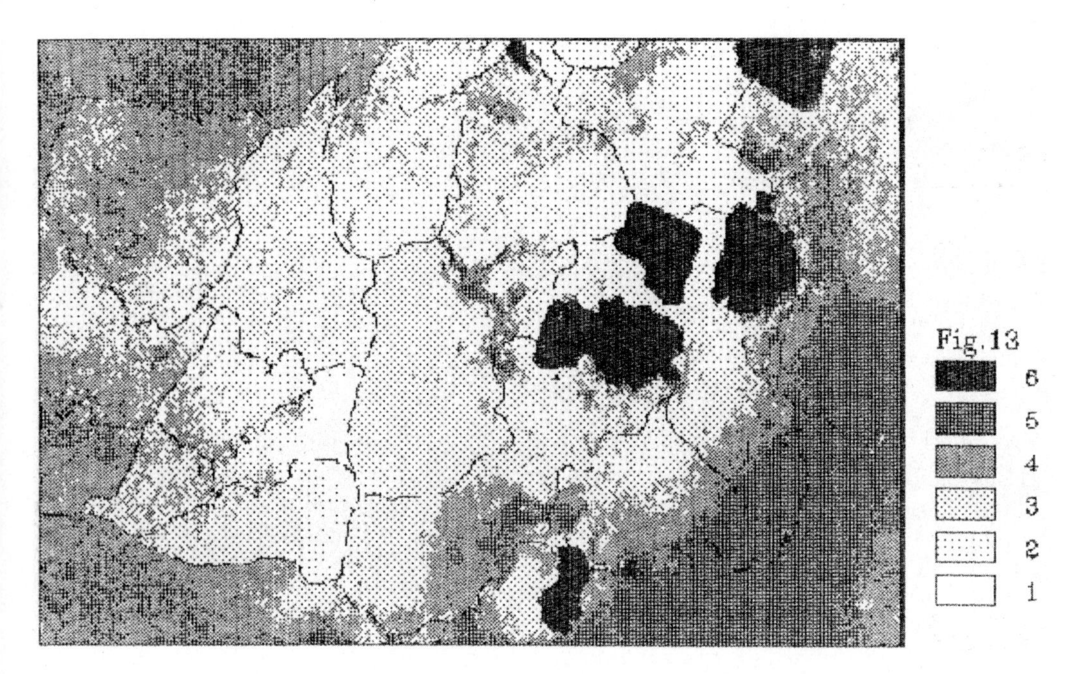

Fig. 13. A Woody biomass supply/demand model indicating surplus and deficit areas within the Weredas of southern Shewa. The annual consumption was assumed to be 1 ton/capita. The woody biomass resource available for consumption was assumed to be 10% of the standing woody biomass. The area covered corresponds to the 1:250 000 topographical sheet EMA3, sheet NB 37-2, Hos'aina (167 x 115 km). The shades correspond to: 1= deficit 20-50, 2= deficit 0-20, 3= surplus 0-20, 4=surplus 20-60, 5= surplus >60 ton/ha, year.

Detailed information about inter-Wereda supply-demand patterns calls for data with a higher resolution and more computer laborious methods as described below.

6.7.3. Modelling on the local levels.

All huts (from air photos and 1:50 000 topographical sheet) and villages were digitized assuming that each hut, or minor clusters of huts, corresponded to one houshold. The average household size for individual Weredas was obtained from the 1984 census data. Approx. 9 000 huts were digitized within the Wilbareg 1:50 000 sheet assuming 4.7 persons/hut. The household size used for the Mertule Mariam area was 4.2 persons.

Different population, supply and demand models, developed at the Lund University Remote Sensing Laboratory, were applied on the data sets to generate e.g.

- population densities/pixel (25 x 25 m), ha and square km (Fig. 14a-c).
- the sum of the supply and the sum of the demand within cells of optional size (e.g. ha and square km) in regular grid nets covering the complete areas (Fig. 15a-c).
- the supply-demand differences of the mentioned grid nets (Fig. 16a-b).
- population pressure/potential exploitation models describing the number of people who could reach every pixel if they were allowed to walk a specified distance from home (Fig. 17).
- the fuelwood surplus or possible deficit of individual households and villages, within a specified walking distance from home, taking the fuelwood accessibility (walking distance) and possible conflicts (several villages or households have access to the same limited fuelwood resource) into account.

This type of modelling can become extremely complex (and computer time consuming) depending on the consumption and conflict rules set up.

6.8. Soil erosion

6.8.1. The Universal Soil Loss Equation.

Detailed assessments of average soil erosion can only be carried out through field observations during long time series. However, a lot of research has been undertaken to establish mathematical relationships between various landscape features and the rate of soil erosion. The most well known example is perhaps the Universal Soil Loss Equation (USLE) (Hudson 1981, Kirkby and Morgan 1980,

Fig.14a. Population density/pixel (25 x 25 m) and major drainage in the surroundings of Mertule Maryam. Every dot equals one household (4.2 persons). The population density in Mertule Maryam is 2.7 persons/pixel. The area covered is 5 by 5 km. The huts were mapped through air photo interpretation and the data was retrieved from the 1:20 000 scale Gojam Case data base.

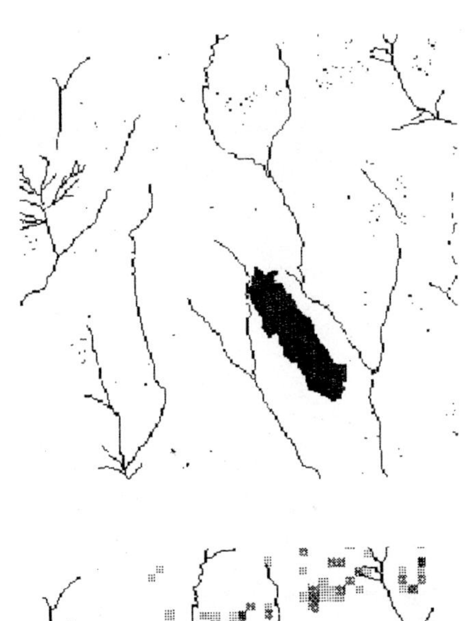

Fig.14b. Population density/ha.

The shades correspond to:

0:	0	INH/HA
1:	4	-"-
2:	8	-"-
3:	12	-"-
4:	16	-"-
5:	20	-"-
6:	20 - 50	-"-
7:	50 - 70	-"-
8:	> 70	-"-

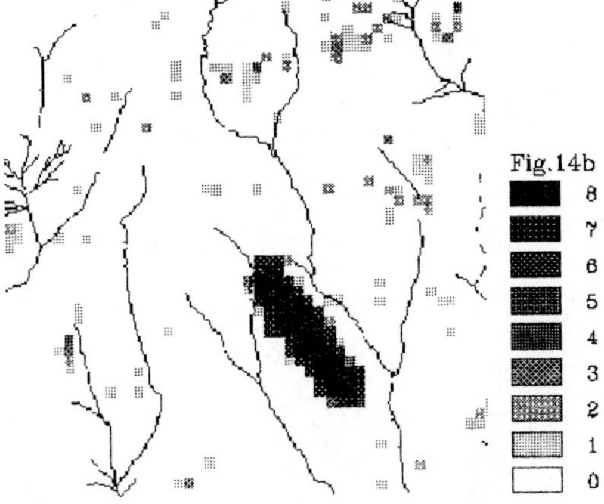

Fig.14c. Population density /km².

The shades correspond to:

0:	25 m contours
1:	0 INH/KM²
2:	20 - 40
3:	40 - 60
4:	60 - 80
5:	80 - 10
6:	100 - 150
7:	150 - 800
8:	800 - 1500
9:	1500 - 2500
10:	> 2500

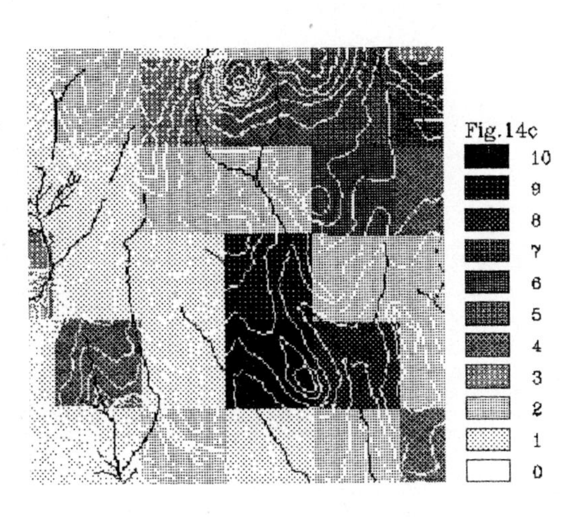

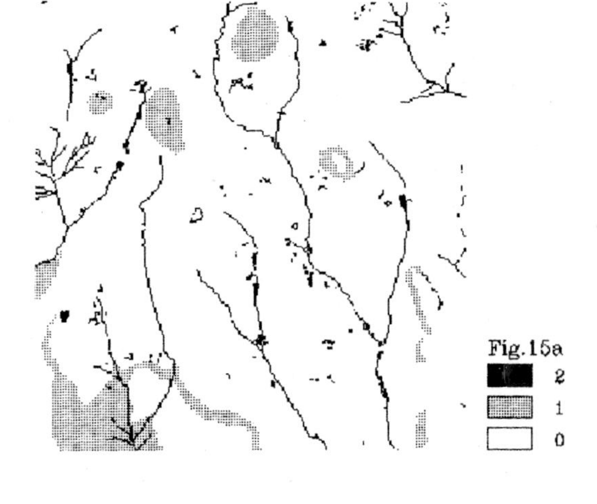

Fig. 15a. Distribution of Bushland and Eucalyptus in the vicinity of Mertule Maryam in 1985 according to air photo interpretation (25 x 25 m pixels). The area covered is 5 by 5 km.

The shades correspond to:

0: Cultivated land
1: Bushland
2: Eucalyptus trees and stands

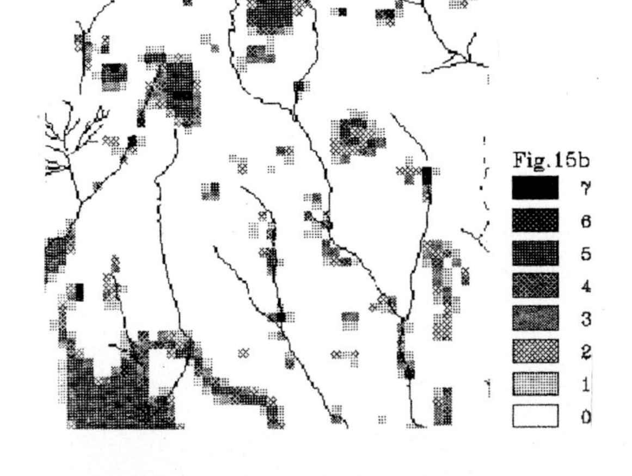

Fig. 15b. Woody biomass in ton/ha.

The shades correspond to:

0:	Cultivated land	
1:	1 - 25	ton/ha
2:	25 - 50	
3:	50 - 75	
4:	75 - 100	
5:	100 - 125	
6:	125 - 150	
7:	> 150	

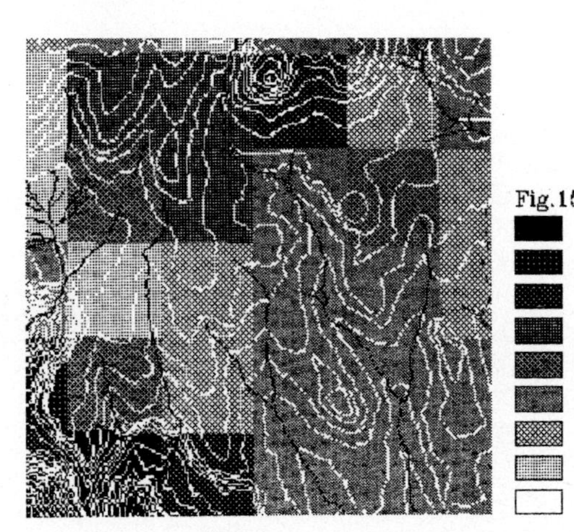

Fig. 15c. Woody biomass in ton/km².

The shades correspond to:

0:	25 m contours	
1:	0	ton/km²
2:	1 - 500	
3:	500 - 1000	
4:	1000 - 1500	
5:	1500 - 2000	
6:	2000 - 3000	
7:	3000 - 5000	
8:	> 5000	

Fig. 16a. The woody biomass supply - demand difference summed up for 1 ha squares in the vicinity of Mertule Maryam. The consumption was assumed to be 1 ton/capita and year and the resource available was assumed to be 10% of the standing woody biomass (the annual growth).
The shades correspond to:
1: Deficit < 75 ton/ha
2: Deficit 50 - 75
3: Deficit 25 - 50
4: Deficit 0 - 25
5: Balance -
6: Surplus 0 - 25
7: Surplus > 25

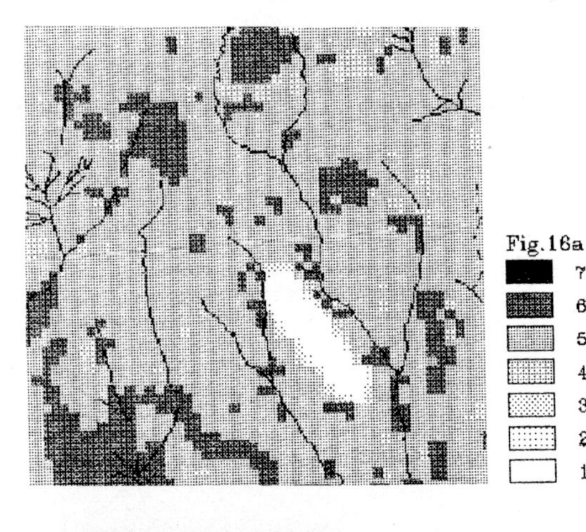

Fig. 16b. The woody biomass supply - demand difference summed up for 1 km squares in the vicinity of Mertule Maryam.
The shades correspond to:

1: Deficit < 100 ton/km²
2: Deficit 50 - 100
3: Deficit 0 - 50
4: Balance -
5: Surplus 0 - 50
6: Surplus 50 - 100
7: Surplus > 100

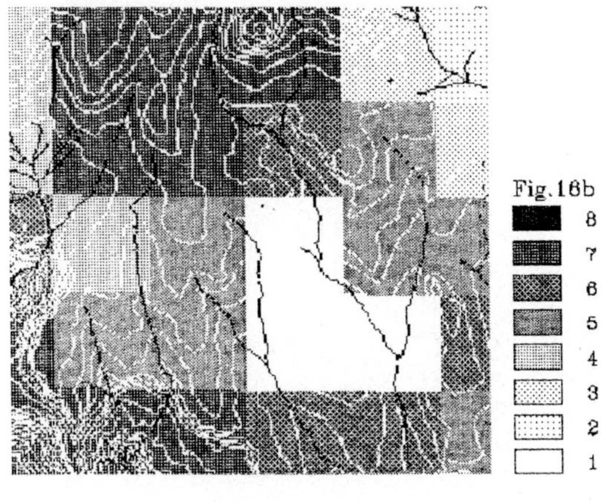

Fig.17. Population pressure plot indicating the number of people that can reach every pixel (25 x 25 m) if they are allowed to walk 1 km from home.
The shades correspond to:

1: Contours, 25 m
2: < 10 people/pixel
3: 10 - 30
4: 30 - 60
5: 60 - 90
6: 90 - 120
7: 120 - 150
8: 150 - 200
9: > 200

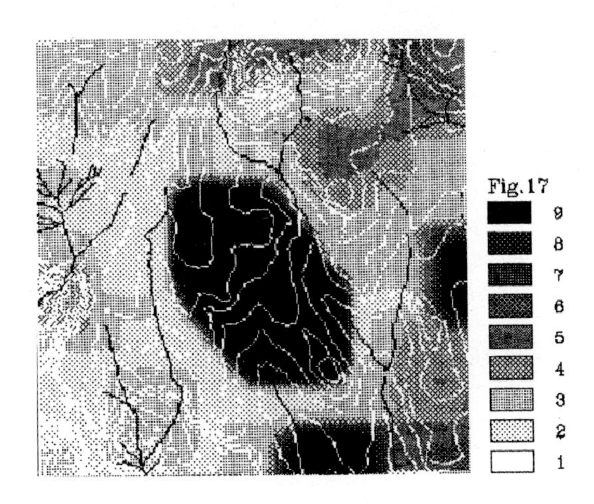

Wischmeier and Smith 1978). The model, primarily designed as a convenient aid for soil conservation planning, was based on empirical research and statistical analysis of data from field experiment stations to allow for soil erosion prediction in the United States for given circumstances. The model has been subsequently adjusted for field applications in many other countries.

It should be noted that the model was developed as a method to predict average annual soil loss from interrill (splash and sheet) and rill erosion only (Kirkby and Morgan 1980). Gully erosion is the most spectacular type of interfluvial erosion, ultimately leading to the creation of badlands. However, interrill and rill erosion are probably the most important in terms of total soil loss.

The model was adapted for Ethiopian conditions through empirical research during the 1980's by the Soil Conservation Research Project, Community Forests and Soil Conservation Development Department, NRCDMD (Hurni 1984, 1985). The model and its factors are summarized in Table IV.

It is obvoius from the table that the land cover, the precipitation characteristics and the topography (slope length and gradient) are the three major environmental factors controlling soil erosion in the order mentioned. The importance of the land cover, assessed through remote sensing, is outstanding.

Soil conservation measures are often concentrated on a modification of the topographical factors as it is difficult to implement a change in the traditional land cover. A reduction of the slope gradient from 20% to 5% through terracing, including a reduction of slope length from eg. 40 m to 10 m, yields a 10 times reduction in the annual soil loss according to USLE.

The Universal Soil Loss Equation was used in this study in a new approach to soil erosion estimation for the assessment of soil conservation needs on national, regional and local levels.

6.8.2. USLE factor estimations.

The following approach was used to assess the USLE environmental factors on a pixel by pixel basis (Cf. Table IV. and Fig. 3).

- Rainfall erosivity: All tabulated monthly precipitation records for 157 rainfall stations (the majority of the Ethiopian stations), including latitude, longitude and height above sea level were fed into the information system. The mean annual precipitation for each station was computed.

Table IV. The Universal Soil Loss Equation (USLE) adapted for Ethiopia. Please notice the relative importance of the different factors.The table was modified from Hurni (1985).

THE UNIVERSAL SOIL LOSS EQUATION (USLE) ADAPTED FOR ETHIOPIA

EQUATION: SOIL LOSS = R * K * L * S * C * P (Tons/ha year)

R: **RAINFALL EROSIVITY** — **Importance: 28**

Annual rainfall (mm):	100	200	400	800	1200	1600	2000	2400
Annual factor R :	48	104	217	441	666	890	1115	1340

K: **SOIL ERODIBILITY** — **Importance: 2**

Soil colour :	black	brown	red	yellow
Factor K :	0.15	0.20	0.25	0.30

L: **SLOPE LENGTH** — **Importance: 8**

Length (m) :	5	10	20	40	80	160	240	320
Factor L :	0.5	0.7	1.0	1.4	1.9	2.7	3.2	3.8

S: **SLOPE GRADIENT** — **Importance: 12**

Slope (%) :	5	10	15	20	30	40	50	60
Factor S :	0.4	1.0	1.6	2.2	3.0	3.8	4.3	4.8

C: **LAND COVER** — **Importance: 1000**

Dense forest:	0.001	Dense grass:	0.01
Other forest:	0.01-0.05	Degraded grass:	0.05
Badlands hard:	0.05	Fallow hard:	0.05
Badlands soft:	0.40	Fallow ploughed:	0.60
Sorghum, maize:	0.10	Ethiopian tef:	0.25
Cereals, pulses:	0.15	Continuous fallow:	1.00

P: **MANAGEMENT FACTOR** — **Importance: 2**

Ploughing up & down:	1.0	Ploughing on contour:	0.9
Strip cropping:	0.8	Intercropping:	0.8
Applying mulch:	0.6	Dense intercropping:	0.7
Stone cover 80%:	0.5	Stone cover 40%:	0.8

Source: Wischmeier and Smith (1978)
Adaptions: R correlation: Hurni (1985)
K values: Bono and Seiler (1983,1984), Weigel(1985)
S extrapolation: Hurni (1982)

Multiple correlation and stepwise multiple regression analysis of the national data set did not reveal any strong relationships between annual precipitation, height above sea level, latitude or longtitude. Therefore, the complete national pixel by pixel coverage was generated through linear interpolation without considering any elevation data (Fig 18).

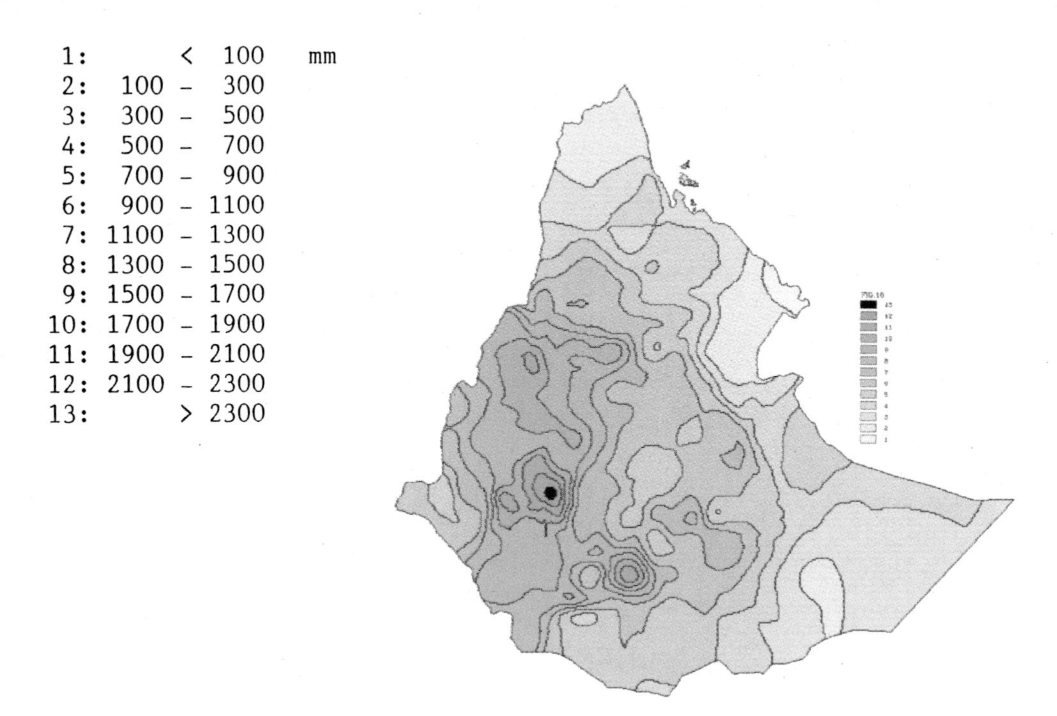

Fig. 18. The mean annual precipitation of Ethiopia. The plot was generated through a linear interpolation of 157 rainfall stations. The original data set pixel size is 2 x 2 km and the rainfall resolution is 10 mm.

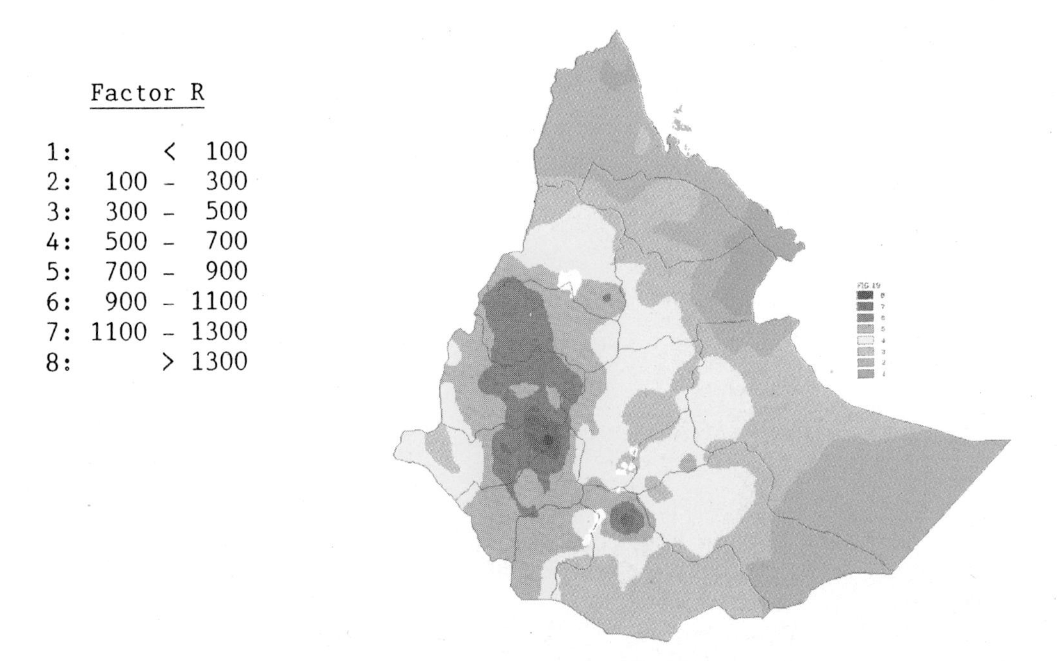

Fig. 19. Rainfall erosivity distribution of Ethiopia. The R-factor was derived as a function of the mean annual precipitation.

The data in Table IV was used to establish the linear regression describing the R-factor as a function of the mean annual precipitation. The resulting equation

$$\text{Factor R} = -8.12 + 0.562 \times \text{Mean annual precipitation} \quad (r^2=1.00)$$

was applied to the national data set to generate a national rainfall erosivity information layer (Fig. 19). Data sets in appropriate scales were created to fit into the data base.

- Soil erodability: The K values were obtained by digitizing existing soil maps (1:1 million), including the K values, generated by the FAO assistance to the land use planning project at the Land Use Planning Department (FAO 1984).

- The topographical factors: The contour lines of the topographical maps were scanned and vectorized through a computerized procedure. Each contour was assigned its correct height value and each data set created was interpolated for the generation of digital elevation models in appropriate scales. Knowing the altitude and location of each pixel in the complete grid nets, software was developed to calculate the gradient of each pixel. The data in Table IV was used to establish the linear regression describing the S-factor as a function of slope. The resulting equation

$$\text{Factor S} = 0.344 + 0.0798 \times \text{Slope (\%)}, \quad (r^2=0.97)$$

was applied to the gradient information layers.

Knowing the gradient of each pixel, all mountain tops were identified by the program and a filter was created calculating the shortest distance from every pixel to its top assignment. Conflict pixels were assigned to the top with the highest height difference. During a second run, all pixel distance values along the shortest way to the top were replaced by the maximum pixel distance value along that path. The model was finally smoothed by applying a mean value filter. The data in Table IV was used to establish the linear regression describing the L-factor as a function of the slope length. The resulting equation was applied to the slope length information layers.

$$\text{Factor L} = 0.799 + 0.0101 \times \text{Length (m)}, \quad (r^2=0.95)$$

Every pixel value in the final slope length information layers represents an approximation of the total water run-off length (from the top of its assigned mountain down to the nearest flat area or valley bottom) corresponding to the meaning and conventional use of the USLE slope length factor. Replacing the "shortest distance to the top" with the "shortest distance, perpendicular to the contours, to a water divide pixel" is a more accurate approach to the slope length assessment. Such an approach is under development.

The slope length estimation procedure is not applicable on the regional level because of the limited geometric resolution selected (pixel size= 1 ha). The length was set to 100 m in every pixel.

The accuracy of the topographical modelling depends on the resolution of the contour data as well as on the pixel size selected. Ditches, contour bunds etc. diverting the surface runoff and changing the length of the overland water flow, cannot be accounted for even if 2 m contour data were available. The problem can be solved on the local level through air photo interpretation, or field surveying, for the identification and mapping of ditches, bunds and minor terraces to be included in the slope length modelling.

- Land cover: The land cover was obtained through the analysis of remotely sensed data (Landsat MSS, TM, SPOT and air photos depending on the scale wanted) as described under section 6.5. "Land use/land cover assessment".

The C-factor values for land cover classes not included in Table IV, e.g. Acacia woodlands, classified into different canopy cover classes, were assessed using Table III as a guideline.

- Management factor: Each land use/land cover class was assigned a P-factor value based on the general field experience and using Table IV as a guideline.

The factor estimation procedure is illustrated in Fig. 20a-i.

6.8.3. Erosion modelling and assessment of conservation needs.

The Universal Soil Loss Equation was applied on the generated data sets on the different planning levels (Cf. Fig. 1 and 3). The result was compared with the tolerable soil erosion presented by Hurni (1985) for the agroclimatic zones of Ethiopia (11 zones, tolerable erosion 2-18 tons/ha, year). The difference was categorized into a number of soil conservation need classes, expressed as the soil loss reduction needed to reach an acceptable level (Fig 21).

By changing the different USLE factors the effect of different soil conservation measures can be modelled and displayed for any drainage basin or administrative unit. Appropriate soil conservation techniques can be selected and factors such as e.g. bunding, grass strip and terrace spacing can be assessed to reach a specified soil loss reduction. This information can be used for a first approximation of the total investments needed for labour, equipment, transports etc. starting on the national/regional level, going into more detailed levels depending on political priorities and decisions.

6.8.4. Severely denuded lands.

Severely denuded lands (large gully patterns, badlands and exposed bedrock) were recognised in the satellite data through a visual pattern interpretation combined with computer identification of vegetation free areas (Fig. 22 a-d). No multitemporal data were included in this study for an accurate identification of severely denuded lands (areas lacking a green vegetation cover during the dry season as well as at the end of the rainy season).

6.9. Result presentation

The major purpose of the information system approach is not to mass produce maps, describing static characteristics of the landscape, but to analyse and model dynamic landscape characteristics, derived from the continuously updated multipurpose data base, for defined planning purposes. The results are normally displayed and studied on graphic screens and the majority of them are never printed. Sometimes, however, hard copies of optional information layers and scales are needed in a limited number.

The enclosed colour maps and 3-dimensional models were generated by means of an Applicon Ink Jet Colour Plotter applying a graphic/cartographic software package to the data base. Black and white maps, tables and diagrams were generated by means of conventional laser and matrix printers. The maps can be generated as transparencies to be superimposed on optional topographical map sheets.

Whenever a large edition is needed, the computer output (the computer colour separated sheets for colour printing) has to be sent to the press for conventional printing.

Fig. 20a. An example of a high resolution mean annual precipitation plot covering the surroundings of Mertule Maryam. The area covered is 5 x 5 km and the pixel size is 20 by 20 m. The plot is a detail of one of the Gojam local level data sets. The shades correspond to:

1: < 1080 mm
2: 1080 - 1090
3: 1090 - 1100
4: 1100 - 1110
5: 1110 - 1120
6: 1120 - 1130
7: > 1130

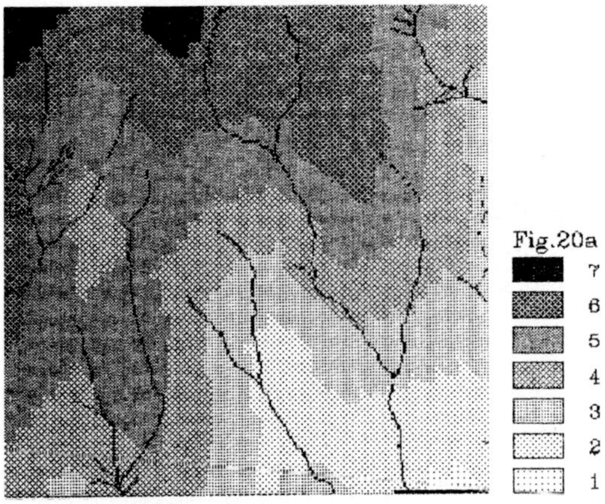

Fig.20a
7
6
5
4
3
2
1

Fig. 20b. Erosivity plot. The shades correspond to:

Annual factor R

1: < 590
2: 590 - 600
3: 600 - 610
4: 610 - 620
5: > 620

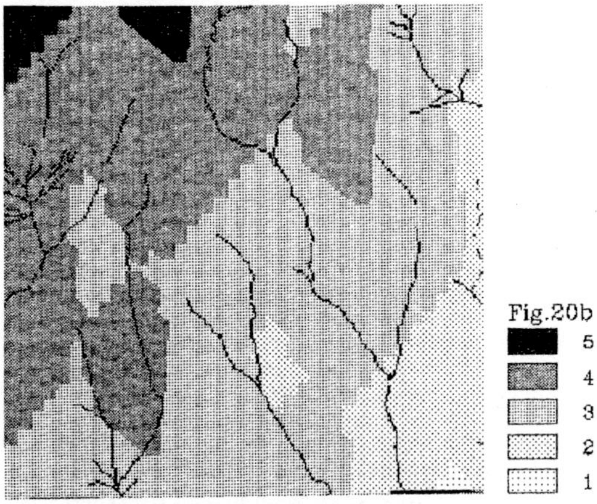

Fig.20b
5
4
3
2
1

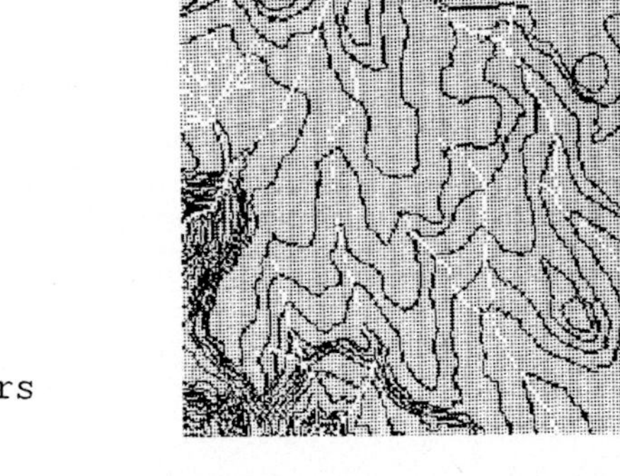

Fig.20c.Elevation contours (25 m).

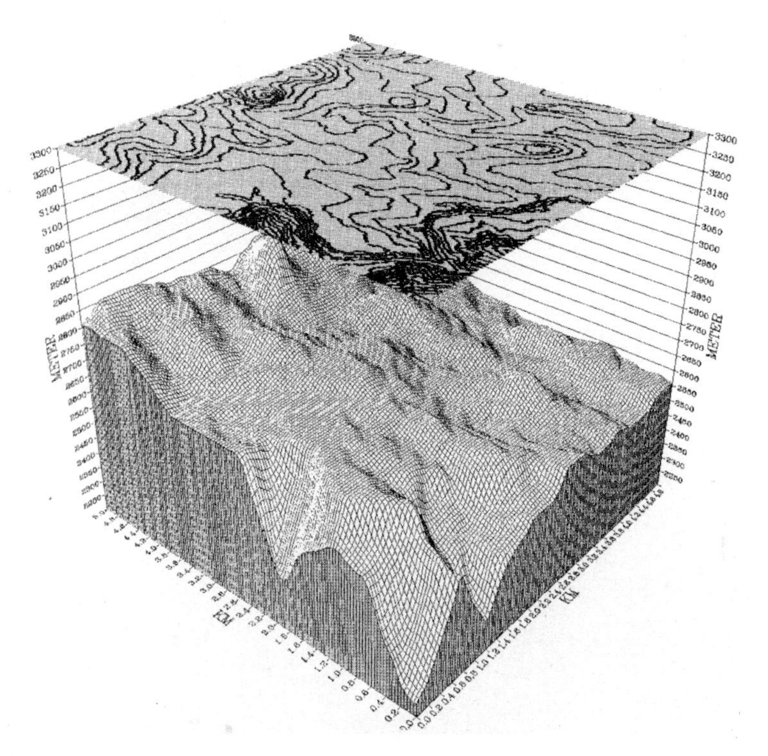

Fig. 20d. An elevation model obtained from digitized contours.

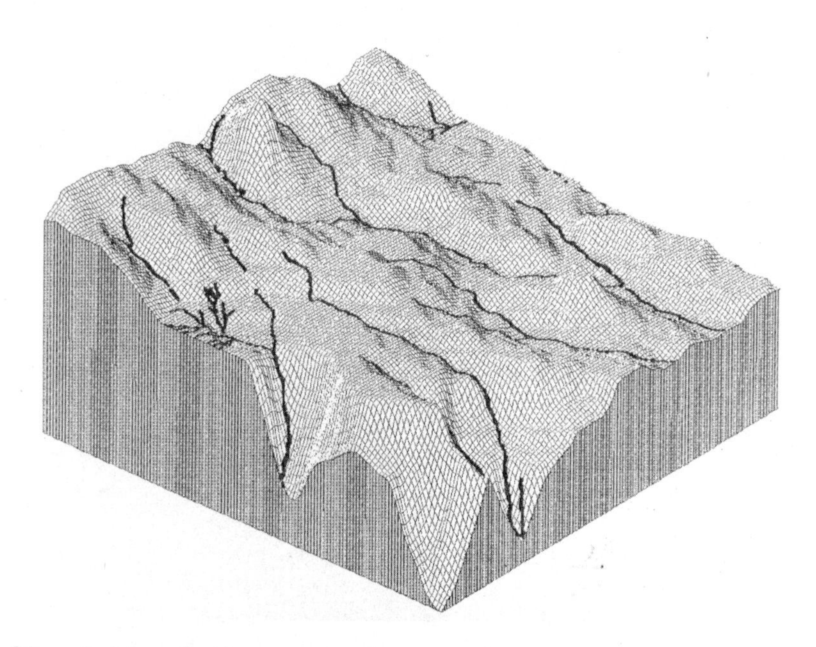

Fig. 20e. Optional themes can be superimposed on the elevation model. The slope length factor (L) and the slope gradient factor (S) can be estimated.

Fig. 20f. Slope gradient factor (S).

The shades correspond to:

1: < 0.5
2: 0.5 - 1.0
3: 1.0 - 2.0
4: 2.0 - 3.0
5: 3.0 - 4.0
6: > 4.0

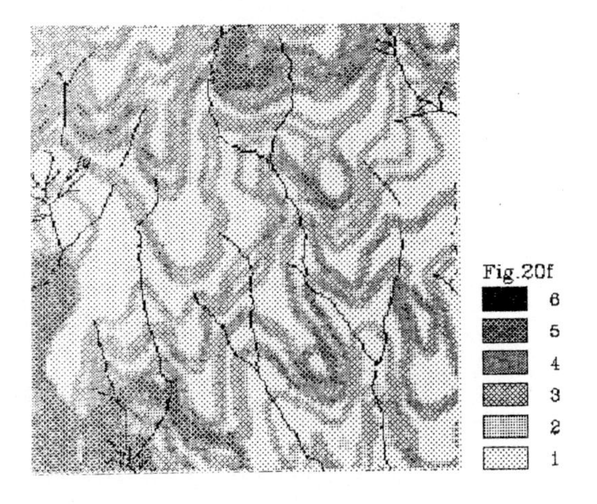

Fig. 20g. Land cover.

The shades correspond to:

1: Cultivated land
2: Eucalyptus stands
3: Bushland
4: Badland
5: Meadow
6: Major drainage

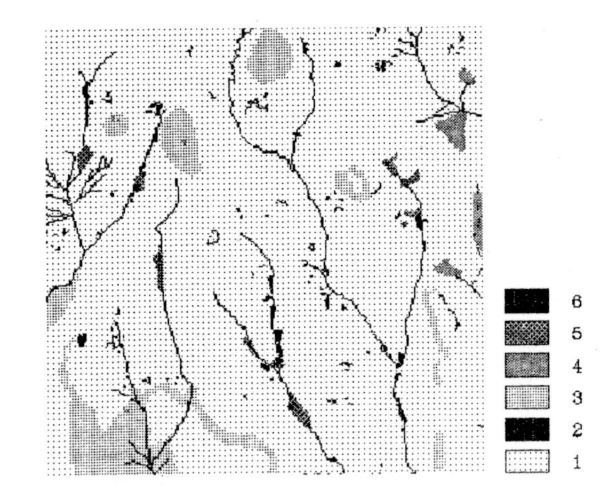

Fig. 20h. Soil loss.

The shades correspond to:

1: < 10 ton/ha, year
2: 10 - 20
3: 20 - 30
4: 30 - 40
5: 40 - 50
6: 50 - 100
7: 100 - 150
8: > 150

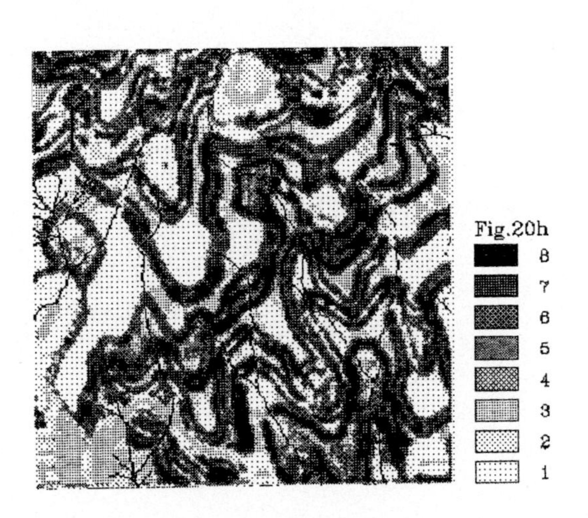

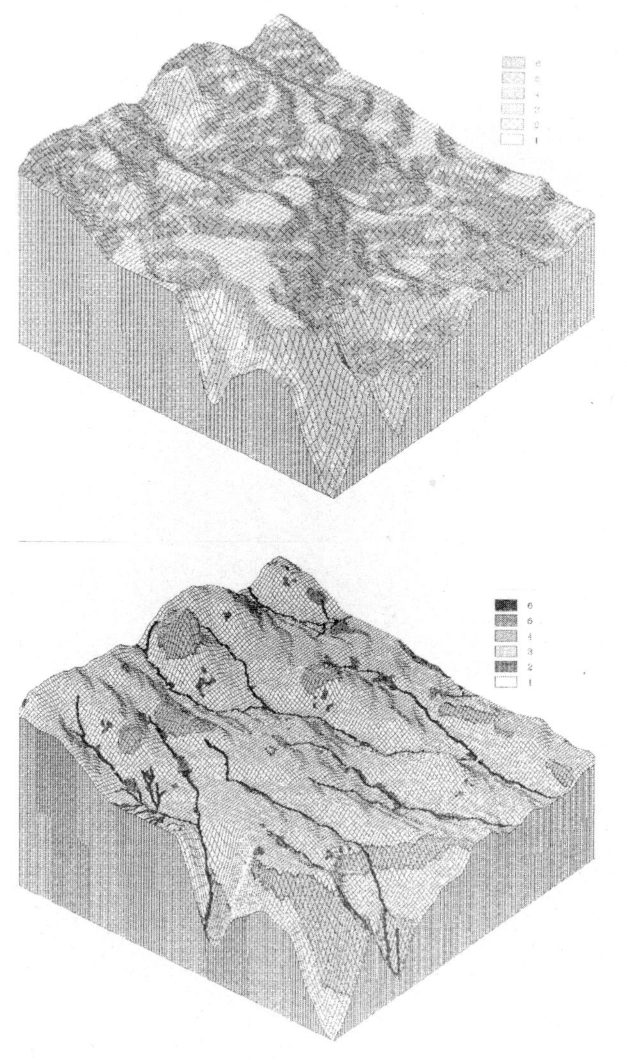

Fig. 21. The terrain model ABOVE illustrates the soil loss reduction needed to reach a tolerable soil erosion level (8 ton/ha, year). The terrain model BELOW describes the land cover (Cf. Fig. 20g). Please notice the bushland influence on the soil loss.

Soil loss reduction:

1: < 1 ton/ha, year
2: 1 - 10
3: 10 - 20
4: 20 - 50
5: 50 - 100
6: > 100

Land cover:

1: Cultivated land
2: Eucalyptus stands
3: Bushland
4: Badland
5: Meadow
6: Major drainage

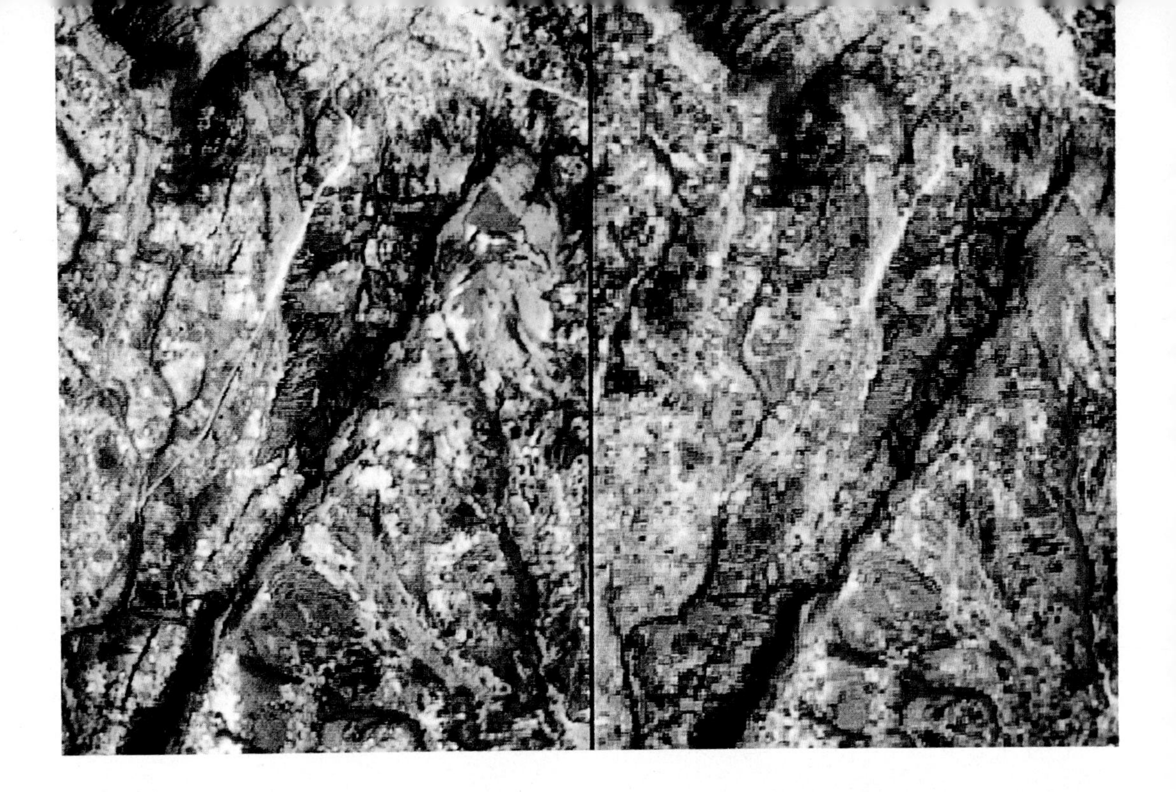

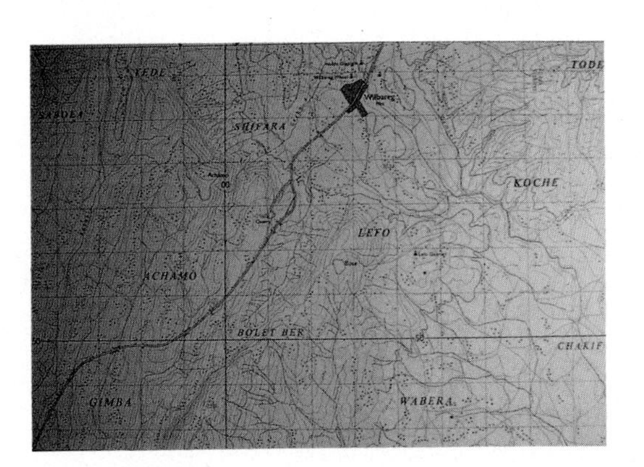

Fig. 22a. SPOT (top left) and Landsat TM (top right) false colour composites covering a severely eroded area just south of Wilbareg. The SPOT pixel size is 20 m and the TM pixel size is 30 m. Dense vegetation cover is indicated in red. Blue and grey colours indicate barren soils and bedrock. Roads, individual houses, canyons, small tree stands, meadows and cultivated fields can be identified. The frame covers 7 by 5 km.

Fig. 22b. The 1:50 000 topographical map coverage of the same area (bottom left).

Fig. 22c. Denuded lands east of the road from Wilbareg to Hosaina.

7. RESULTS

7.1. Introduction

More than 600 information layers, holding raw data, intermediate results, inventory results and models, were generated on different scales and under different assumptions for the two major areas under study. To keep inside the economical frames of the feasibility study , just a few of them were prepared for printing to be included in this report. All of them are of course available in digital form.

The most important results are summarized below. They should not be considered as the final and long lasting results of a multipurpose inventory, but rather as examples of possible and temporary inputs to the planning process. Input assumptions, models, themes and class limits can always be modified/replaced when they are considered to be unrealistic and the information layer in question can be regenerated. It implies that the system allows for a continuous improvement of all results and models, with time and through local experience/expert input, in an operational environment.

7.2. The Shewa Case

7.2.1. Land use/land cover.

The land use/land cover classes were selected with the main purpose of fitting into the USLE modelling, considering the classes available in Table IV.

Information generated at the 1:250 000 generalization level in the form of maps (information layers) and statistics:

- Canopy cover/pixel.
- Classified (categorized) canopy cover: 0-1, 2-10, 11-20, 21-30, 31-50, 51-70, 71-100 %.
- Administrative boundaries: Awrajas (2) and Weredas (17).
- Infrastructure: major roads and villages
- Elevation contours: 50 m and 100 m.
- Land cover, land use, USLE land cover and management factors according to Table V.
- Major drainage.
- Landscape models (3-dimensional models) in different perspectives.
- Tables and histograms summarizing the land use/land cover classes coverage and proportions per province (Awraja), county (Wereda) and map sheet.

Table V. The Shewa Case land cover classes on the 1:250 000 generalization level.

Land use/land cover	Canopy Cover	USLE C-factor	USLE P-factor
Badlands	0-2 %	0.35	1.0
Trad. cult./grazing	3-30	0.10	0.7
Woodland/grazing	31-50	0.05	0.7
Dense woodland	51-70	0.01	0.6
Forest	71-100	0.001	0.5
Intensive cult.	0-50	0.25	0.9
Wetlands	-	-	-
Moderately cult./grazing	31-50	0.08	0.7

Information generated at the 1:50 000 generalization level in the form of maps (information layers) and statistics:

- Canopy cover/pixel.
- Classified canopy cover: 0-5, 6-10, 11-20, 21-30 and 100 %.
- Infrastructure: major roads, villages, and huts.
- Elevation contours: 20 m.
- Land cover, land use, USLE land cover and management factors according to Table VI.
- Major drainage.
- Landscape models (3-dimensional models) in different perspectives.
- Tables and histograms summarizing the land use/land cover classes coverage and proportions per km^2 and map sheet.

Table VI. The Shewa Case land cover classes on the 1:50 000 generalization level.

Land use/land cover	Canopy cover	USLE C-factor	USLE P-factor
Badland	0 - 5 %	0.35	1.0
Trad.cult./grazing	6 - 20	0.10	0.7
Woodland	20 - 30	0.03	0.5
Forest	100	0.001	0.5
Eucalyptus stands	100	0.01	0.5
Intensive cultivation	0 - 5	0.25	0.9
Water	-	-	-
Wetlands/marshes	-	-	-

7.2.2. Woody biomass supply/demand modelling.

Information generated at the 1:250 000 generalization level in the form of maps (information layers) and statistics:

-Standing woody biomass/pixel.
-Classified standing woody biomass: 0-1, 2-10, 11-20, 21-30, 31-50, 51-70, >70 ton/ha.
-Woody biomass annual growth.
-Population density in each community (Wereda) excluding the large lakes from the Wereda area calculations.
-Consumption models.
-Supply/demand models.
-Landscape models (3-dimensional models) in varying perspectives with different models superimposed.
-Tables on actual surplus and deficit for each Wereda under different consumption and supply assumptions.

Information generated at the 1:50 000 generalization level in the form of maps (information layers) and statistics:

-Standing woody biomass/pixel.
-Classified standing woody biomass: 0-0.1, 0.1-0.5, 0.5-1.0, 1.0-1.5, 1.5-2.5, >2.5 ton/pixel (25 x 25 m).
-Summaries of standing woody biomass in ton/ha and ton/km^2.
-Woody biomass annual growth.
-Population distribution/pixel.
-Population distribution and density/ha and per km^2.
-Population pressure/exploitation models.
-Consumption models.
-Supply/demand (magnitude of surplus and deficit) summed up per ha and per km^2.
-Pixel by pixel supply/demand descriptions and models under different assumptions allowing people to collect fuelwood within specified distances from home.
-Landscape models (3-dimensional models) in varying perspectives with different models superimposed.

7.2.3. Soil erosion modelling

Information generated at the 1:250 000 and 1:50 000 generalization levels in the form of maps (information layers):

-Annual precipitation/pixel.
-Digital elevation models including pixel gradient and slope length (slope length=100 m in every pixel on the 1:250 000 level).
-Information layers describing all USLE factors on a pixel by pixel basis (Cf. Table IV,V and VI).

Fig. 23. A Landsat MSS false colour mosaic covering the major part of the Shewa study area corresponding to the 1:250 000 topographical sheet EMA3, NB37-2, Hos'aina. Each frame indicate a 1:50 000 sheet (27.5 x 27.5 km). The top left frame corresponds to the Wilbareg sheet.

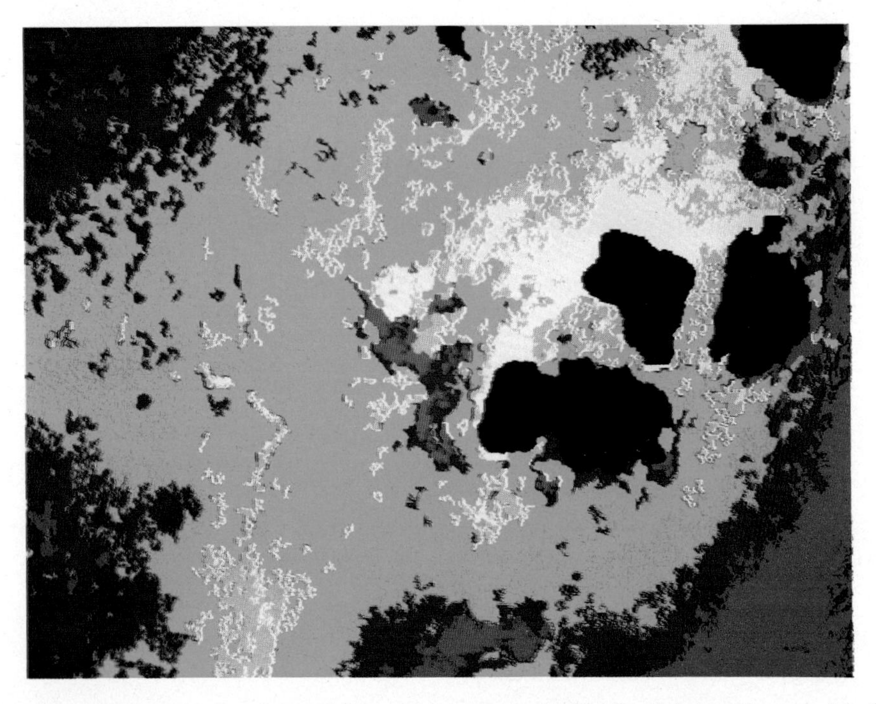

Fig. 24. Canopy cover: Yellow=0-1, Orange=2-10, Light Green=11-20, Dark Green=21-30, Brown=31-50, Red=50-100%.

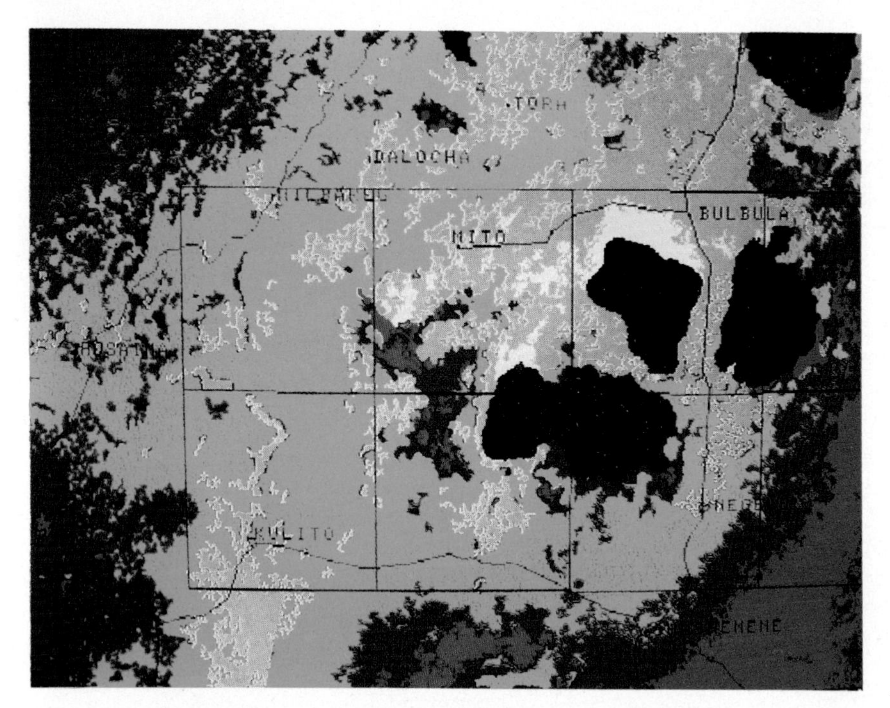

Fig. 25. Standing woody biomass: Yellow=0-1, Orange=2-10, Light Green=11-20, Dark Green=21-30, Brown=31-50, Red= >50 ton/ha.

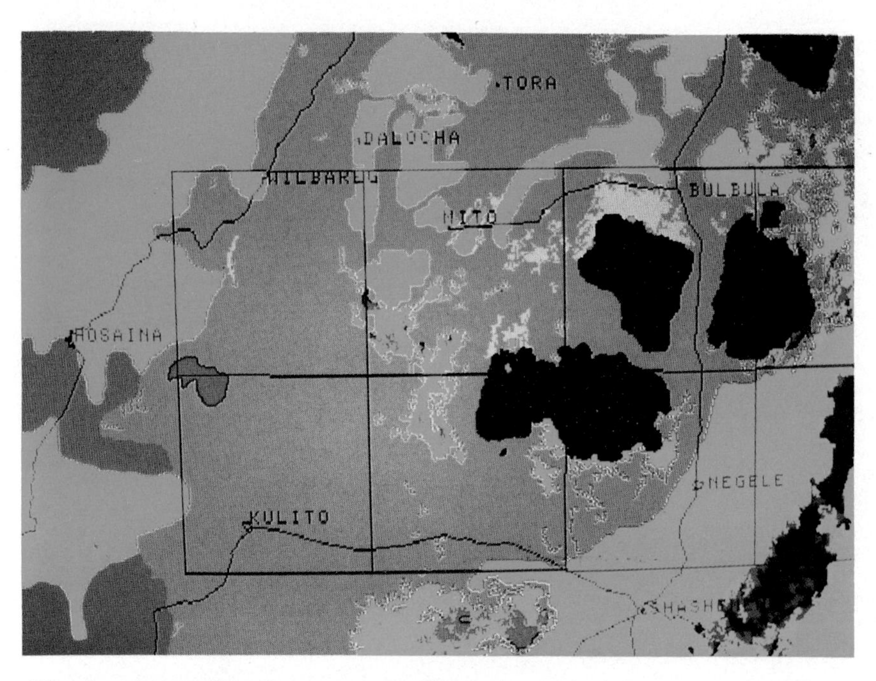

Fig. 26. Land use/land cover: Yellow=degraded land, Orange= traditional cult./grazing, Light Green=wood & bushland/grazing, Dark green=dense woodland, Pink= Moderately cult./grazing, Light Blue=Intensive cult., Brown=forest, Blue=wetlands.

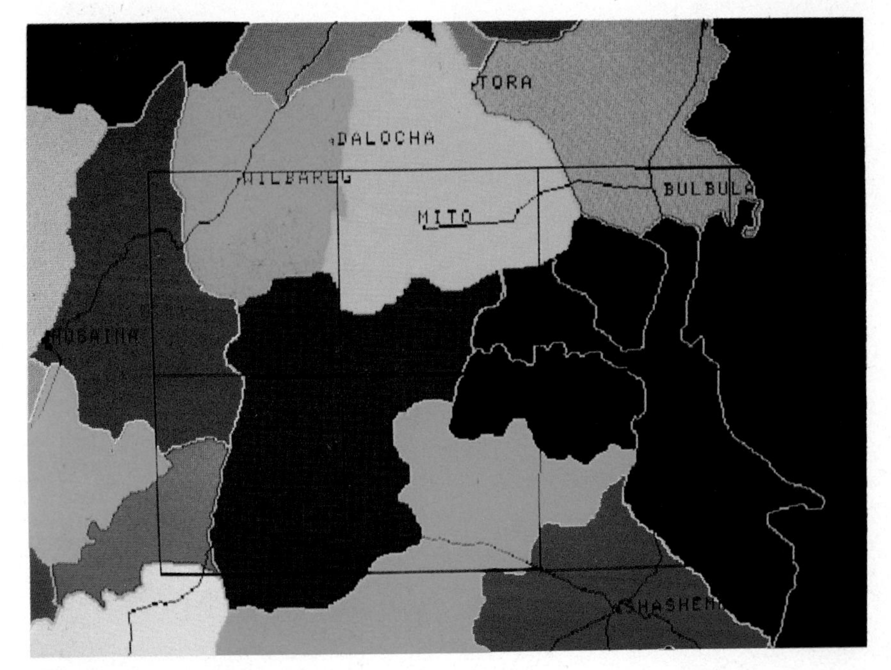

Fig. 27. Administrative units (Weredas) holding population data. (Cf. Table IIIa and IIIb).

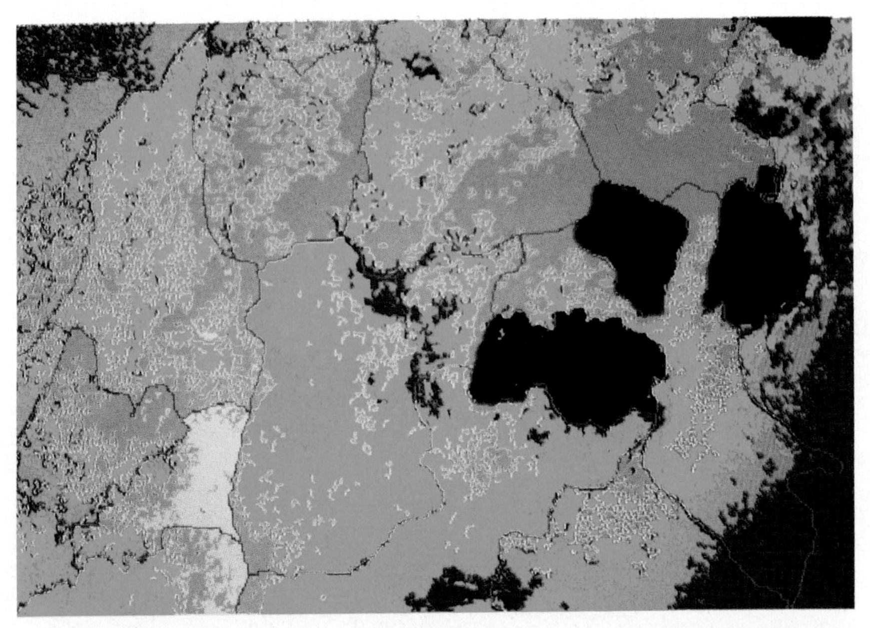

Fig. 28. A model of woody biomass surplus and decificit areas assuming that the population is evenly distributed within each Wereda. Yellow=deficit 20-50, Orange=deficit 0-20, Light Green=surplus 0-20, Dark Green=surplus 20-60, Brown=surplus >60 ton/ha, year (Cf. Fig. 13 & Table IIIa-IIIb).

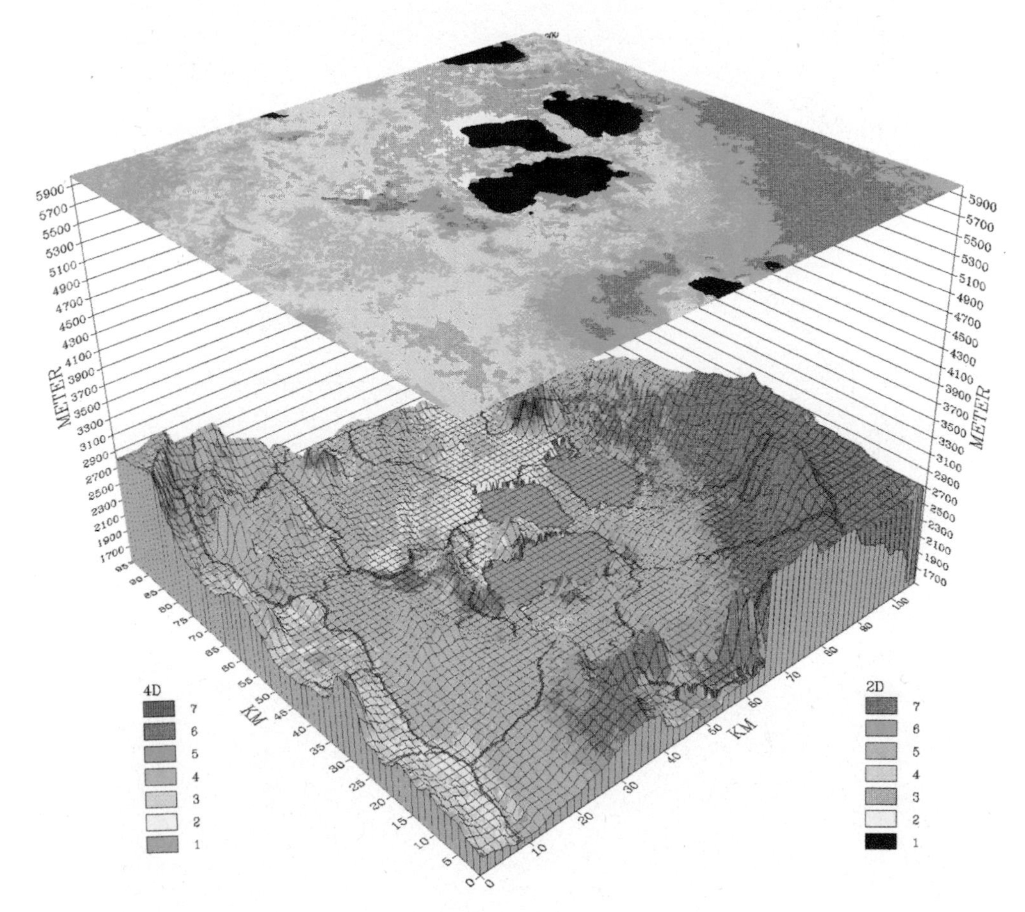

Fig. 29. An example of a woody biomass supply/demand terrain model indicating surplus and deficit areas within the Weredas of southern Shewa. The annual consumption was assumed to be 1 ton/capita. The woody biomass resource available for consumption was assumed to be 20% of the standing woody biomass. The annual consumption was assumed to be 0.6 ton/capita. The canopy cover is superimposed on the 2D-plot.

Supply-demand (4D):

1: Lakes
2: Deficit 20 - 0 ton/ha
3: Surplus 1 - 20
4: Surplus 20 - 60
5: Surplus 60 - 200
6: Surplus > 200
7: Wereda boundaries

Canopy cover (2D):

1: Lakes
2: 0 - 2 %
3: 3 - 10
4: 11 - 20
5: 21 - 30
6: 31 - 50
7: > 50

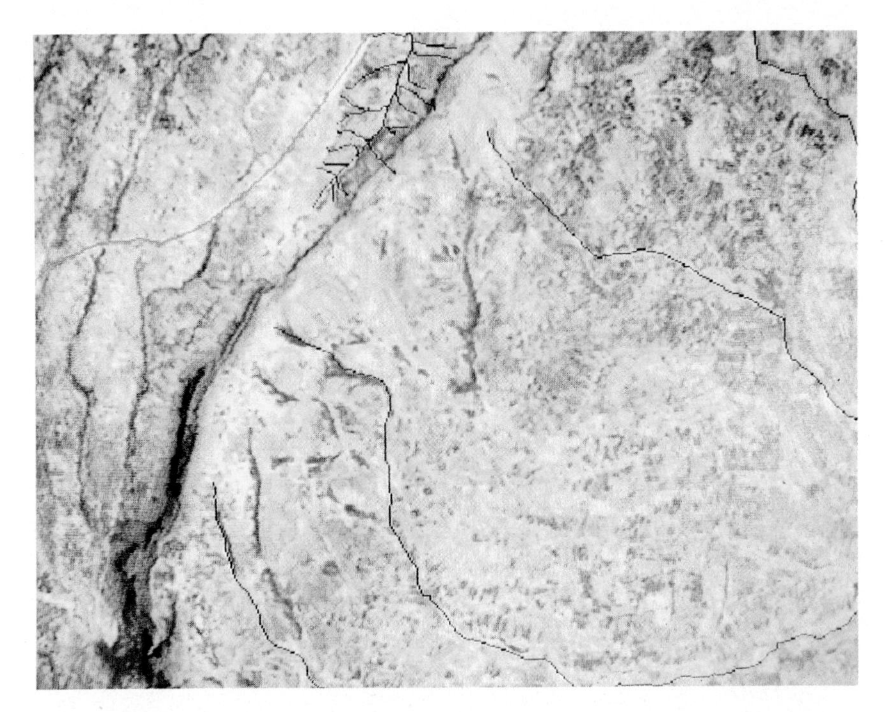

Fig. 30. A detail of a Landsat TM false colour composite covering a 13 by 9 km area just south of Wilbareg (Topo sheet 0738 A3, Wilbareg, 1:50 000). Major roads and drainage are enhanced.

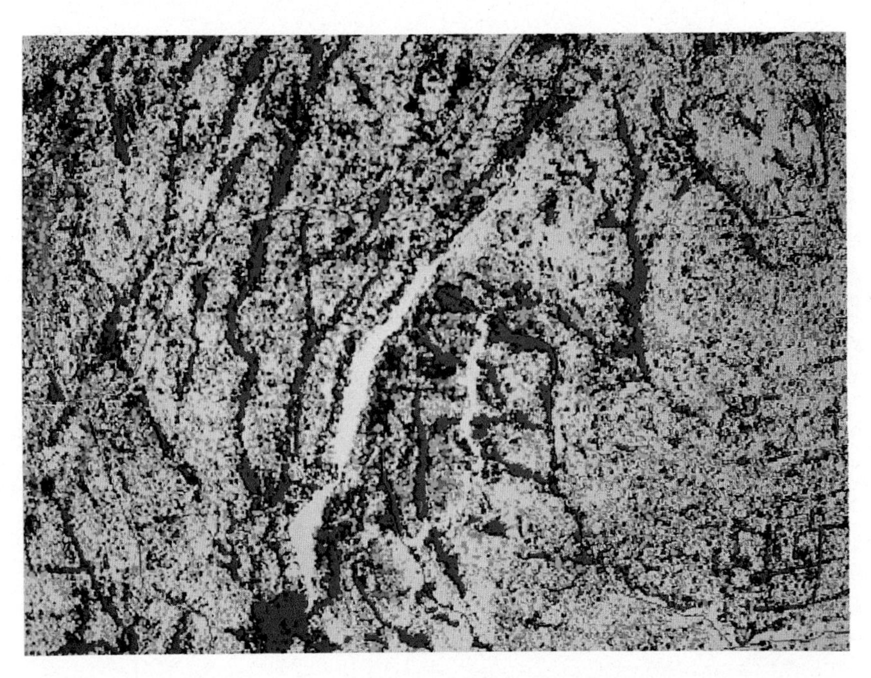

Fig. 31. A standing woody biomass model of the area south of Wilbareg. The pixel size is 25 by 25 m. Yellow=0-1, Orange=1-5, Light green=5-10, Dark green and brown=10-25, Red= >25 ton/pixel.

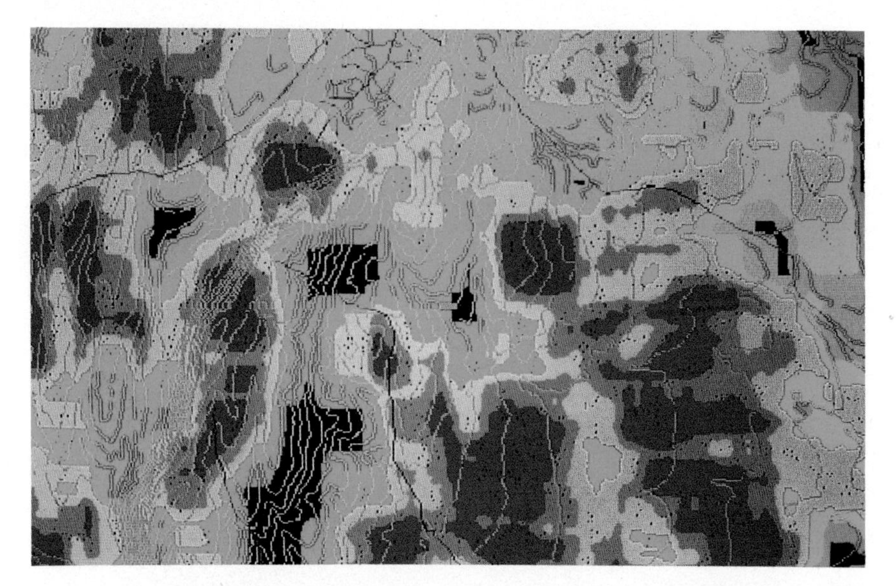

Fig. 32. An example of a population pressure/exploitation model illustrating the number of people which can reach every pixel (25 by 25 m) if they were allowed to walk 1 km from home. Black=0, Dark blue=1-30, Light blue=30-60, Dark blue=60-90, Yellow=90-120, Orange=120-200, Red= >200 people/pixel.

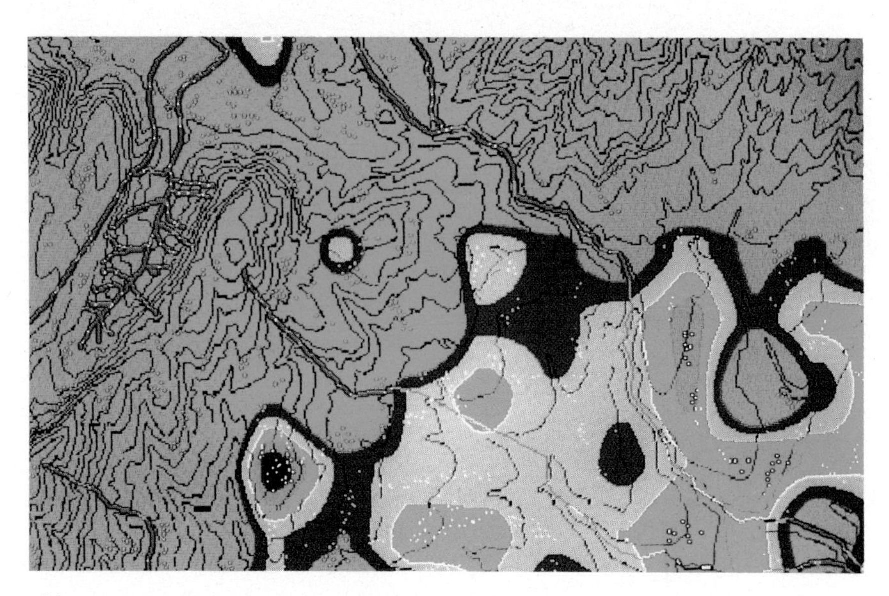

Fig. 33. An example of a woody biomass supply-demand model on the detailed level (Wilbareg data set) summing up the difference between supply and demand per km^2. The annual consumption was assumed to be 1 ton/capita. The woody biomass resource available for consumption was assumed to be 10% of the standing woody biomass. Black=deficit > 50, Dark blue=deficit 50-25, Light blue=deficit 25-0, Yellow=balance, Orange=surplus 0-25, Brown=surplus 25-50, Green= surplus >50 ton/km^2, year.

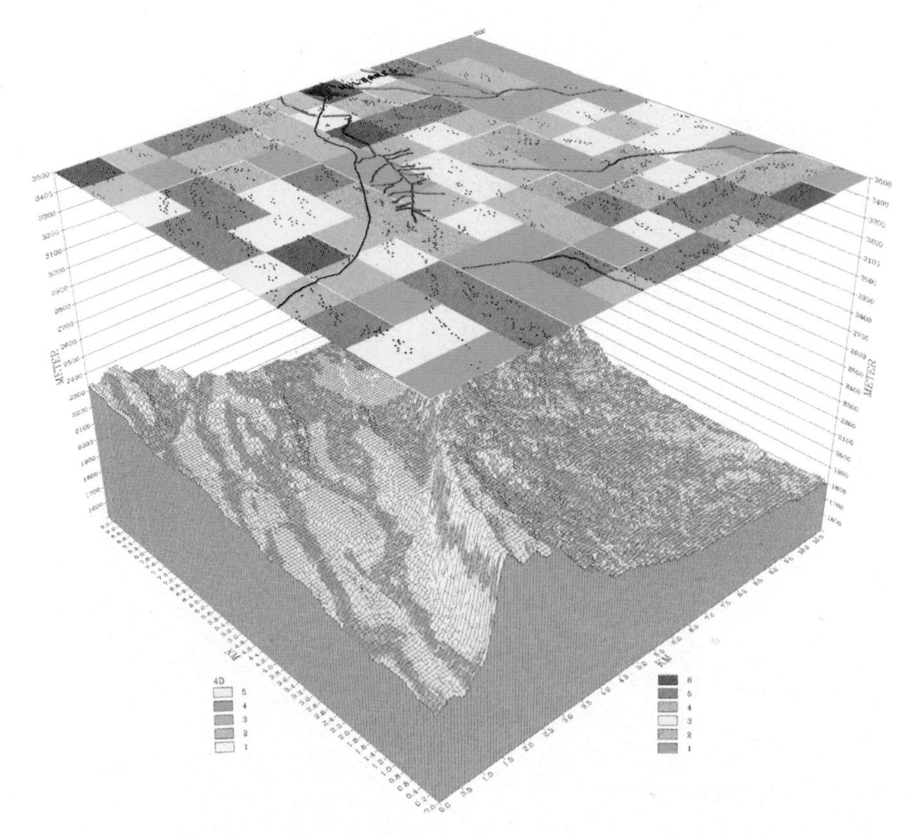

Fig. 34. A terrain model illustrating the land use/land cover distribution (factor C in USLE, Cf. Table VI) on the local level south of Wilbareg. The population density, huts, tracks and the major drainage pattern are superimposed on the 2D-plot.

Land use land cover (4D):	Population density (2D):
1: Denuded land	1: 0 - 25 inh/km²
2: Trad.cult./grazing	2: 25 - 50
3: Woodland	3: 50 - 75
4: Forest	4: 75 - 100
5: Intensive cultivation	5: 100 - 150, 6: >150

-Average soil loss in ton/ha.
-Soil conservation models describing the soil loss reduction needed (ton/ha) to reach a tolerable soil erosion value.
-Landscape models (3-dimensional models) in varying perspectives with different soil loss related models superimposed.

7.3. The Gojam Case

7.3.1. Land use/land cover.

Information generated at the 1:250 000 generalization level in the form of maps (information layers) and statistics:

-Canopy cover/pixel.
-Classified canopy cover: 0, 1-10, 11-20, 21-40, 100 %.
-Community forest (Eucalyptus) distribution.
-Administrative boundaries: Awrajas (4) and Weredas (16).
-Available land/capita and household.
-Agricultural land available/capita and household.
-Infrastructure: major roads and villages.
-Digital elevation model and contours: 50 m and 100 m.
-Landscape models (3-dimensional models) in different perspectives.
-Land cover, land use, USLE land cover and management factors according to Table VII.
-Major drainage.
-Tables and histograms summarizing the land use/land cover classes coverage and proportions per province (Awraja), county (Wereda) and map sheet.

Table VII. The Gojam Case land cover classes on the 1:250 000 generalization level.

Land use/land cover	Canopy Cover	USLE C-factor	USLE P-factor
Int. cultivation	0-1	0.25	0.9
Degraded grassland	1-10	0.05	0.5
Open woodland/grazed	10-20	0.04	0.5
Woodland	20-40	0.01	0.5
Forest	100	0.001	0.5
Eucalyptus stands	100	0.05	0.5
Meadows/grazing,fodder	0-1	0.01	0.5

Information generated at the 1:20 000 and 1:10 000 generalization levels in the form of maps (information layers) and statistics:

- Eucalyptus distribution 1982 and 1985 (individual trees and stands).
- Infrastructure: major roads and tracks, villages, and huts.
- Available land/capita and household.
- Agricultural land available/capita and household.
- Elevation contours: 25 m and 10 m respectively.
- Land cover, land use, according to Table VIII.
- Major drainage.
- Landscape models (3-dimensional models) in different perspectives.
- Tables and histograms summarizing the land use/land cover classes coverage and proportions per km^2, ha and map sheet.

Table VIII. The Gojam Case land cover on the 1:20 000 and 1:10 000 generalization levels.

Land use/land cover	Canopy Cover	USLE C-factor	USLE P-factor
Int. cultivation	0-1 %	0.25	0.9
Meadows/grazing/fodder	0-1	0.01	0.5
Bushland	>20	0.01	0.5
Forest/Eucalyptus stands	100	0.01	0.5
Badland	0-1	0.35	1.0

7.3.2. Woody biomass supply/demand modelling.

Information generated at the 1:250 000 generalization level in the form of maps (information layers) and statistics:

- Standing woody biomass/pixel.
- Classified standing woody biomass: 0-1, 1-10, 10-30, 30-60, 61-110, 110-150 ton/ha.
- Woody biomass annual growth.
- Population density in each community (Wereda) including and excluding non habitable areas like steep hill slopes and canyons from the Wereda area calculations.
- Consumption models.
- Supply/demand models.

-Landscape models (3-dimensional models) in varying perspectives with different models superimposed.
-Tables on actual surplus and deficit for each Wereda under different consumption and supply assumptions.

Information generated at the 1:20 000 and 1:10 000 generalization levels in the form of maps (information layers) and statistics:

-Standing woody biomass in kg/pixel (20 x 20 m and 10 x 10 m).
-Eucalyptus standing woody biomass in kg/pixel.
-Summaries of standing woody biomass in ton/ha and ton/km^2.
-Woody biomass annual growth.
-Population distribution/pixel (20 x 20 m and 10 x 10 m).
-Population distribution and density/ha and per km^2.
-Population pressure/exploitation models.
-Consumption models.
-Supply/demand (magnitude of surplus and defecit) summed up per ha and per km^2.
-Pixel by pixel supply/demand descriptions and models under different assumptions allowing people to collect fuelwood within specified distances from home.
-Landscape models (3-dimensional models) in varying perspectives with different models superimposed.

7.3.3. Soil erosion modelling.

Information generated at the 1:20 000 and 1:10 000 generalization levels in the form of maps (information layers):

-Annual precipitation/pixel.
-Digital elevation models including pixel gradient and slope length.
-Information layers describing all USLE factors on a pixel by pixel basis (Cf. Table IV, VII and VIII).
-Average soil loss in tons/ha.
-Soil conservation models describing the soil loss reduction needed (tons/ha) to reach a tolerable soil erosion value.
-Landscape models (3-dimensional models) in varying perspectives with different soil loss related models superimposed.

Fig. 35. A Landsat MSS false colour composite (100 m pixels) covering a 53 by 38 km portion of the Gojam Case study area (topographical sheet EMA3, NC 37-6, Debre Mark'os). Please notice Debre Mark'os and its surrounding urban forest (Eucalyptus).

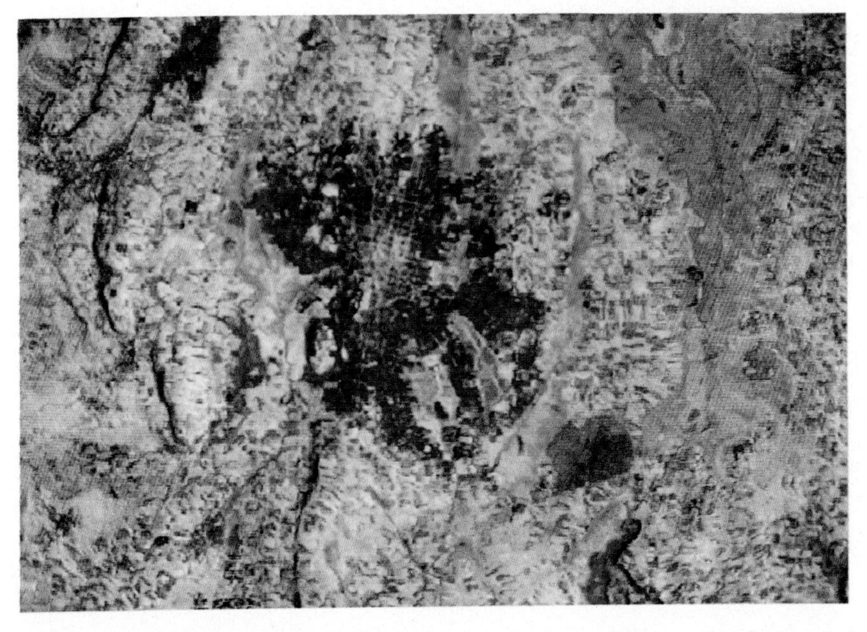

Fig. 36. A Landsat TM false colour composite (30 m pixels) with Debre Mark'os in the center. The Eucalyptus stands can easily be separated from the cultivated fields and meadows.

Fig. 37. A Landsat MSS false colour composite (100 m pixels) covering a portion of the the Gojam Case 1:250 000 data base focusing on Mertule Maryam (Cf. Fig. 11-12). The yellow frames (27.5 by 27.5 km) indicate 1:50 000 sheets. The white frame indicate the coverage of the local and most detailed data bases (1:20 000 and 1:10 000).

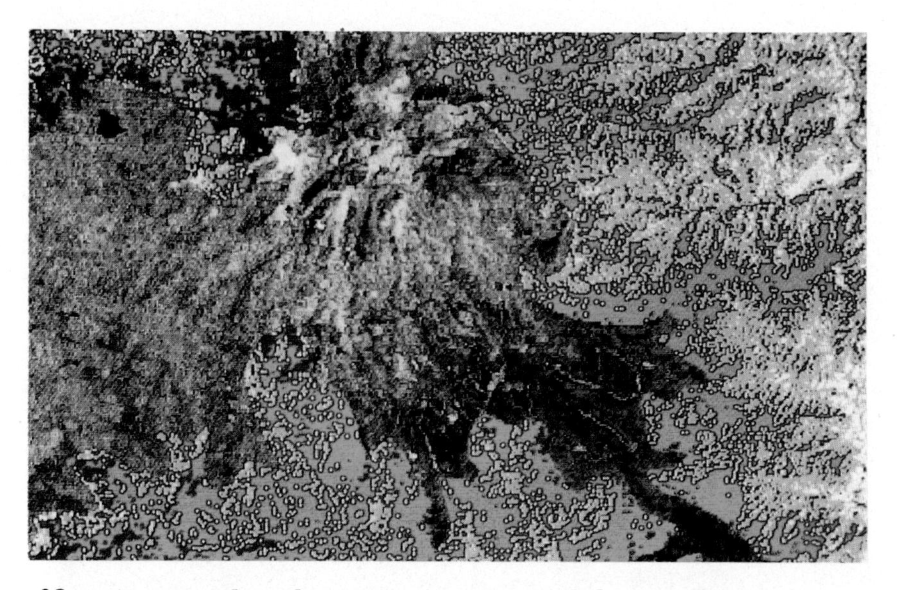

Fig. 38. An example of a canopy cover model superimposed on the Landsat MSS false colour composite (100 m pixels).

Yellow= 1-10%, Orange= 10-20%, Green= 20-40%, Brown= 100% canopy cover. The woodlands are dominated by Acacia spp.. The brown colour corresponds mainly to Eucalyptus community forest stands.

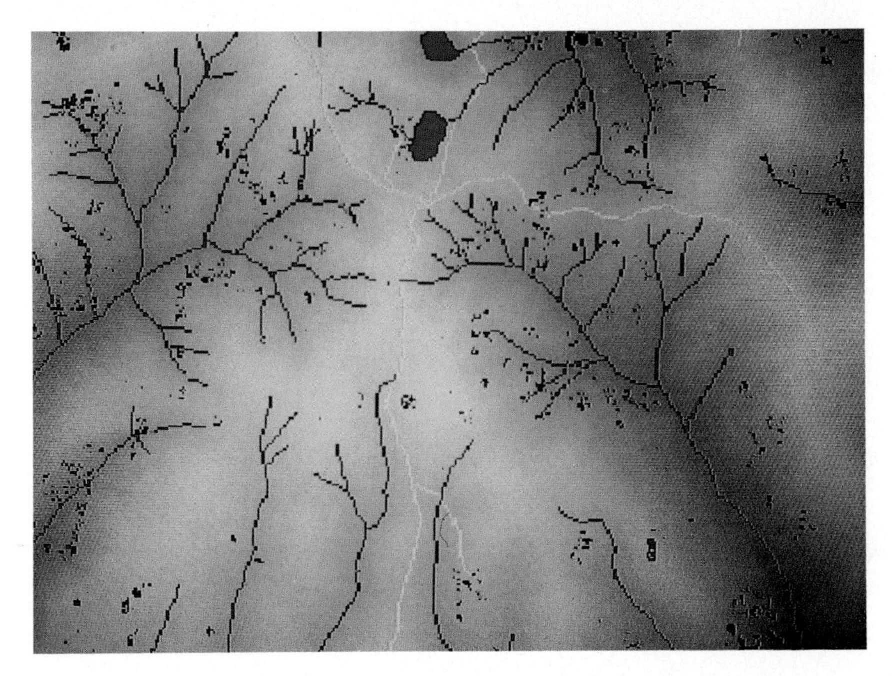

Fig. 39. The distribution of population (huts), Eucalyptus stands, tracks and drainage superimposed on a digital elevation model (pixel size 20 by 20 m). The area covered (10 by 7.5 km) is a portion of the northern part of the Mertule Maryam 1:20 000 data base. Red=huts, Green=Eucalyptus stands, Yellow=tracks, Blue= drainage.

Fig. 40. A portion of a Landsat TM false colour composite (20 m pixels) covering the same area as above.

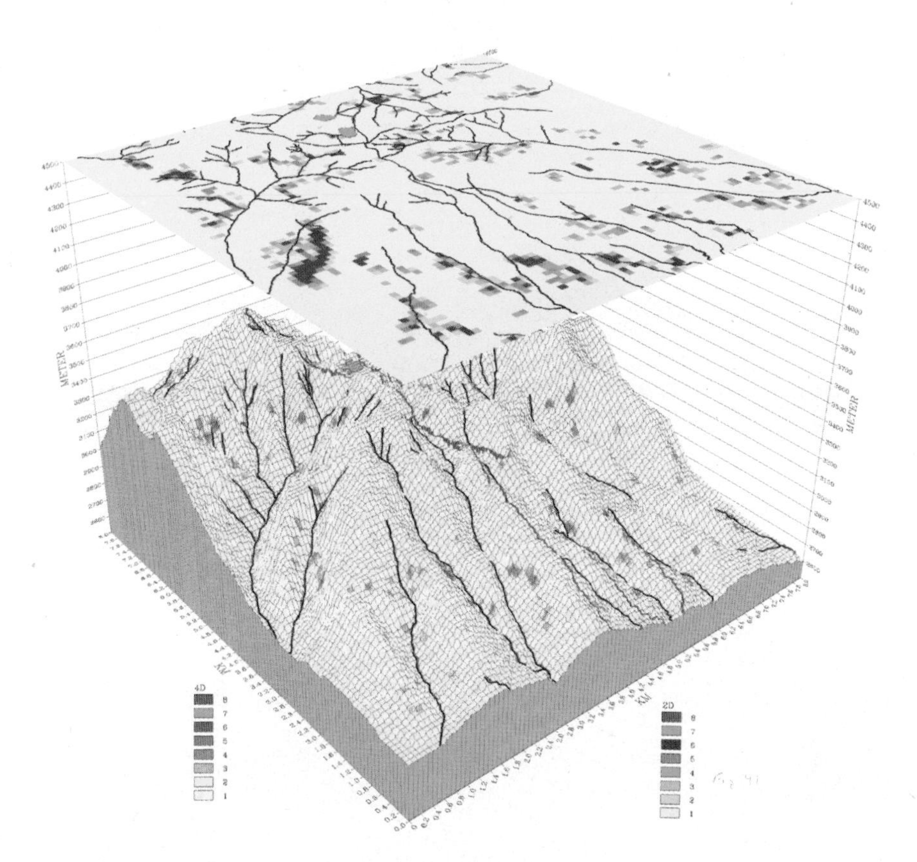

Fig. 41. An example of a terrain model (20 m pixels) describing the population pressure/pixel when people are allowed to move 100 m away from home. A model of the standing woody biomass is superimposed on the 2D-plot. The area covered is a portion of the northern part of the Mertule Maryam 1:20 000 data base (Cf. Fig. 39-40).

Population pressure (4D):

1: 0 inh/pixel
2: 1 - 5
3: 5 - 10
4: 10 - 15
5: 15 - 20
6: > 20
7: Drainage
8: Tracks

Standing woody biomass (2D):

1: 0 ton/ha
2: 1 - 15
3: 15 - 30
4: 30 - 45
5: 45 - 60
6: > 60
7: Drainage
8: Tracks

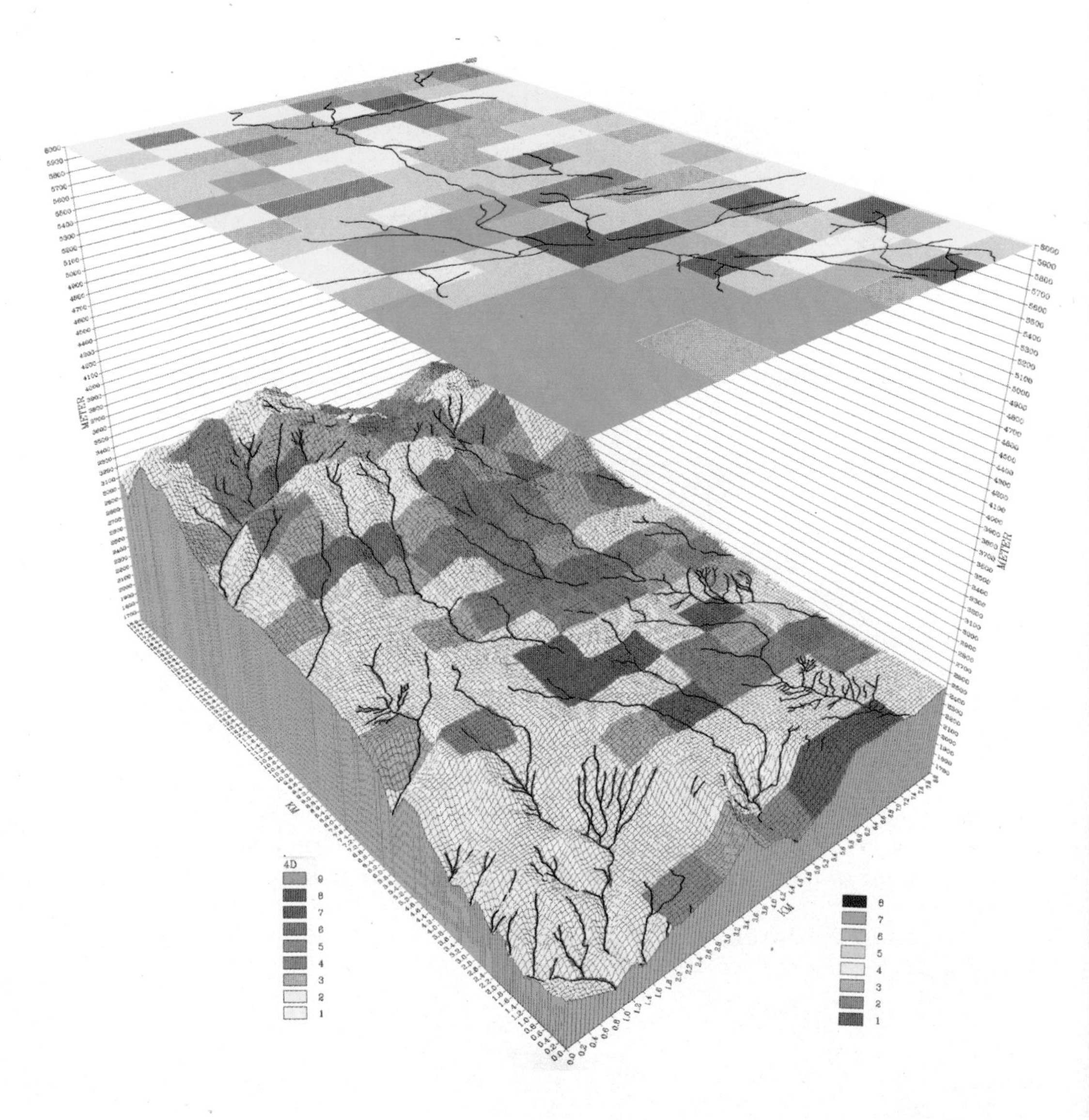

Fig. 42.

Fig. 42. A terrain model (20 m pixels) covering the major area of the Mertule Maryam 1:20 000 data base. An illustration of the population distribution is superimposed on the terrain model. An example of a woody biomass supply-demand model is superimposed on the 2D-plot. The annual per capita consumption was assumed to be 1 ton. The woody biomass resource available for consumption was assumed to be 10% of the standing woody biomass.

Population density (4D):

1:	0	inh/km^2
2:	1 - 20	
3:	20 - 40	
4:	40 - 80	
5:	80 - 100	
6:	100 - 200	
7:	200 - 1000	
8:	> 1000	
9:	Drainage	

Woody biomas supply-demand (2D):

1:	Deficit	> 100	ton/km^2
2:	Deficit	100 - 50	
3:	Deficit	50 - 0	
4:	Balance	-	
5:	Surplus	0 - 50	
6:	Surplus	50 - 100	
7:	Surplus	> 100	
8:	Roads and tracks		

7.4. Integrated planning. A few examples.

In an operational phase the planners are supposed to work in close physical and intellectual interaction with the information system (and its operators) for an optimal use. An operational information and planning system also needs a steady stream of input data from local expertise and from operating field survey teams connected to the information system. Local experts are needed to obtain proper input data for several of the assumptions that have to be made. The field survey teams are needed for a continuous process of model and result calibrations, verifications/modifications as well as for field data collection on the most detailed levels. The possible organisation of a national strategic and multipurpose information and planning system is illustrated in Fig. 43.

A number of questions that may receive an "immediate" (a few minutes - a few days) answer by the information system, are listed below to illustrate some of its capabilities. The examples are based on the assumption that the national/regional information layers, presented in the preceeding chapters, were generated with a national coverage.

-what are the fuelwood and timber resources available within specified distances from all larger villages and towns?

-how many years are needed to deplete the woody biomass resources of the country, of each province (Awraja), of each community (Wereda) for given values on population increase, natural regeneration and possible afforestation efforts?

-assuming it is a political goal that each Awraja should rely on its own woody biomass resources, how many hectares must be afforested to reach a balance in supply-demand within 10 years, 15 years, 20 years? How many seedlings and nurseries are needed/administrative unit? What are the investments needed per region, per province and per community (Awraja)?

-considering the average crop yield and the land available for cultivation/household and for grazing (in relation to the minimum per capita area requirement), is there enough land to afforest (today, tomorrow)?

-if there is not enough land to afforest what is the overpopulation? What are the options? Stimulate migration or import fuel wood from another area? What are the distances, transport needs and costs involved for a fuelwood and timber import from existing woody biomass surplus Awrajas?

-is the woody biomass deficit Awraja suffering from untolerable high soil loss? If so, is it probable that soil conservation measures can increase the crop yield enough to release land for afforestation?

-what are the annual investments needed (labour, tools, total costs)/Awraja or major drainage basin to reduce the soil loss to an acceptable level within 5 years, 10 years, 20 years?

The answers to these and similar questions, when integrated with political goals and priorities, might form the basis for a national and strategic plan on the management and future use of the natural resources. Action priorities can be set up to guide further survey and monitoring investments to cover limited areas with increasingly accurate (and increasingly expensive) information. Questions similar to those listed above can be answered again to set up more detailed plans and priorities, always based on up-to-date information.

The process will continue until an acceptable information level (the management/local level) has been reached for the final decision on where and how to implement a development project and how to monitor its progress in terms of environmental and socio-economic impact.

Change studies and trend analysis can be initiated at any level as soon as historical data (satellite data, air photos, maps, field data) are available in the system.

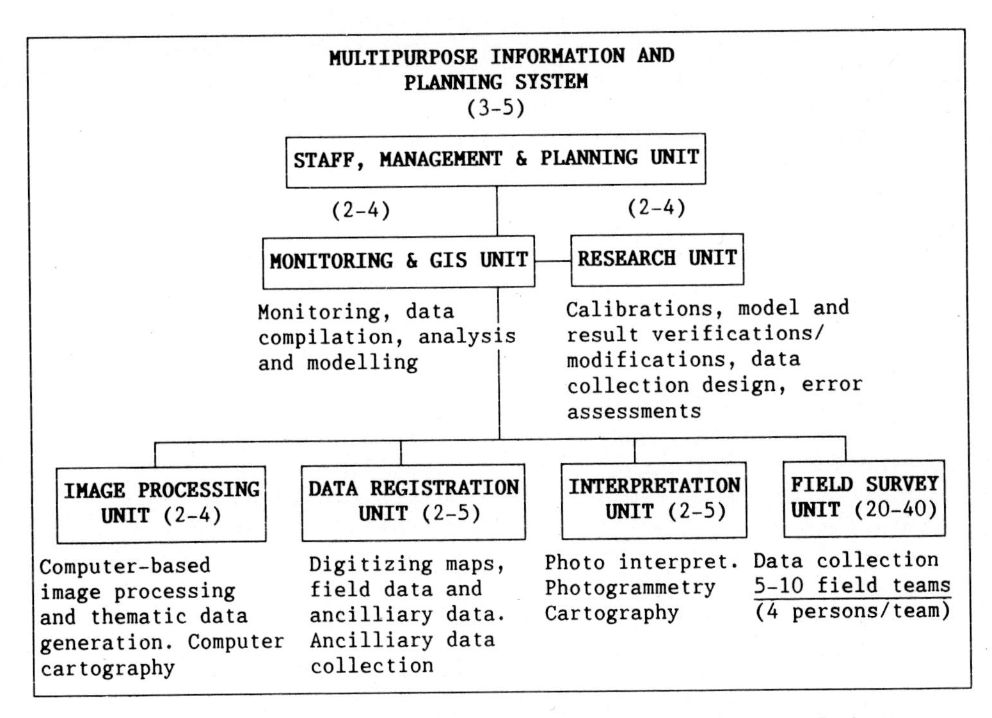

Fig. 43. An indication of essential components of a multipurpose information and planning system. An assessment of the number of qualified people needed to run each component is indicated within brackets.

8. RESTRICTIONS

It is important to stress again that it is an impossible task to acquire a national coverage of accurate and timely information for natural resources planning purposes. It must be understood that the information technology approach illustrated in this study is impaired by errors. New errors are added at every step and on all levels of the data and information flow from the data collection through the different steps of analysis and modelling procedures. The magnitude of the errors involved decrease in size when going from the national to the local level, but they are never absent. Two examples are given below.

-Assume that the standing woody biomass in the Siraro Wereda (124 838 ha) in southern Shewa was overestimated by as little as 40 kg in every 1 ha cell. The total overestimate is almost 5 000 tons for the Wereda, some 32 000 tons for the Awrada and close to two million tons on the national level (corresponding to the fuelwood consumption of 1 million people during 2-4 years).

-When applying the Universal Soil Loss Equation fairly small mistakes can yield large inaccuracies in the resulting soil loss. An erroneous interpretation of satellite data, mixing up dense grass fields with maize and sorghum fields and counting with 100 m slope length instead of 20 m slope length, in a number of 1 ha pixels, creates a 20 times soil loss increase in those pixels, assuming the other four USLE factors have not changed. An average annual soil loss of 20 ton/ha might be acceptable but not 400 ton/ha. Introducing minor errors in the assessment of the other USLE factors will of course result in even more erroneous soil loss estimations.

It implies that the information approach, including the computer based information system, can only yield rough approximations of the distribution of the natural resources in time and space on the national level. It can serve as a guide only, for supply - demand analysis on this level, to set up national strategic plans and priorities to steer future investments into more detailed data collection and analysis. However, a national information system based on guesses is the only realistic alternative.

It should be stressed again that it is essential that a field survey organisation is integrated with the information system to operate on the national as well as on the regional and local levels as indicated in Fig. 43. Field data collection, based on different strategies, will always be needed in a strategic monitoring and planning system for

- remotely sensed data calibrations
- model and result verifications and modifications
- environmental and socio-economic surveys on the most detailed levels.

9. COSTS

The present feasibility study covered more than 40 000 km². The total project cost was approximately 10 US Dollar/km².

The annual costs for an operational national system, including the components indicated in Fig. 43, would be much lower. The cost interval 2 - 6 US Dollar/km² and year is probably a realistic assessment. It implies an annual cost of 2 - 7 million US Dollar for a national multipurpose strategic planning unit in a country like Ethiopia (1.2 million km²) depending on the capacity wanted for the three different information and planning levels.

10. REFERENCES

Ahlcrona,E., 1987: Monitoring the impact of climate and man on land transformation.-A study in arid and semi-arid environment in central Sudan.- (Prel. titel) Doctor thesis in prep. Remote Sensing Lab., Department of Physical Geography, University of Lund.

Andersson,A.,Ekelund,L., 1985: A review of the Ethiopian Topographical Mapping Project Three.- Swedsurvey. 33 pp.

Central Statistics Office, 1985: Population of Weredas, and towns, by sex, and average household size based on the preliminary census results; and population projections by age-sex groups and rural-urban, for total country and regions: 1984-1995.- Census Supplement 1. The Provisional Military Government of Socialist Ethiopia. Office of the National Committee for Central Planning. Central Statistics Office. Addis Ababa. 370 pp.

FAO, 1984a: Assistance to Land-Use Planning. Ethiopia. Agroclimatic resources inventory for land use planning. AG:DP/ETH/78/003. Technical Report 2: Vol III Maps. - The Provisional Military Government of Socialist Ethiopia, Ministry of Agriculture, Land Use Planning and Regulatory Department, UNDP, FAO.

FAO, 1984b: Assistance to Land-Use Planning. Ethiopia. Geomorphology and Soils. AG:DP/ETH/78/003, Field Document 3, maps and Technical Reports.- The Provisional Military Government of Socialist Ethiopia, Ministry of Agriculture, Land Use Planning and Regulatory Department, UNDP, FAO.

Hall-Könyves,K., 1987: Crop studies by means of satellite data.- (Prel. titel) Doctor's thesis in prep. Remote Sensing Laboratory, Department of Physical Geography, University of Lund.

Helldén,U.,Olsson,K., 1982: The potential of Landsat MSS data for wood resources monitoring-A study in arid and semi-arid environment in Kordofan, The Sudan. - Lunds Universitets Naturgeografiska Institution. Rapporter och Notiser Nr. 52, 35 pp.

Howe and Gulick 1980: Fuelwood and other renewable energies in Africa, a progress report on the problem and respons. L'energie dans les communautes rurales des pays du Tiers Monde. C.N.R.S. Domain Universitaire de Bordeaux 33405 Talence. 24 pp. (ref. to in Olsson,K., 1985b).

Hudson, N., 1985: Soil conservation.- Batsford Academic and Educational. London ISBN 0 7134 3521 6, 324 pp.

Hurni,H., 1984: Soil Conservation Research Project, Compilation of Phase I Progress Reports (years 1981, 1982 and 1983) Vol 1-4. UNiversity of Berne, United Nations University & Ministry of Agriculture, Soil and Water Conservation Department, Provisional Military Government of Socialist Ethiopia.

Hurni,H., 1985: Soil Conservation Manual for Ethiopia (First Draft).- Soil Conservation Research Project. Community Forests and Soil Conservation Development Department, Natural Resources Conservation and Development Main Department, Ministry Of Agriculture. Addis Abeba. 86 pp.

Justice,C.O., (editor) 1986: Monitoring the Grasslands of Semi-arid Africa using NOAA-AVHRR Data.-Reprint from the International Journal of Remote Sensing, Vol. 7, No. 11, pp. 1383-1622.

Kir,A., 1984: Assistance to Land-Use Planning. Ethiopia. Forest Resources and Potential for Development. AG:DP/ETH/78/003, Tecnical Report 7.-UNDP,FAO, Rome. 46 pp.

Kirkby,M.J.,Morgan,R.P.C., (editors) 1980: Soil erosion.- John Wiley & Sons. ISBN 0 471 27802 5. 312 pp.

Markham,B.L.,Barker,J.L., 1986: Landsat MSS and TM Post-Calibration Dynamic Ranges, Exoatmospheric Reflectances and At-Satellite Temperatures.- EOSAT. Landsat Technical Notes No.1, August 1986. pp. 3-8.

Olsson,L., 1985: An integrated study of desertification.- Meddelanden från Lunds Universitets Geografiska Institution. Avhandlingar XCVIII. (Dr. thesis). 170 pp.

Olsson,K., 1985a: Fuelwood demand and supply in the Umm Ruwaba/Er Rahad Region in N.Kordofan, The Sudan.-A study based on field data and Landsat MSS information.- Lunds Universitets Naturgeografiska Institution. Rapporter och Notiser Nr. 64., 74 pp.

Olsson,K., 1985b: Remote sensing for fuelwood resources and land degradation studies in Kordofan, the Sudan.- Meddelanden från Lunds Universitets Geografiska Institution. Avhandlingar C (Dr. thesis). 182 pp.

Orgut-Swedforest 1984: Feasibility study.- Community forests, Land use and soil erosion/conservation inventory, Ethiopia. Phase one.-Swedforest Solna, Sweden. 33 pp.

Orgut-Swedforest 1986: Feasibility study on community forests, land use/land cover and soil erosion/conservation inventory in Ethiopia. Phase II. - A debriefing report from Stage I. Addis Abeba, March 1986. 7 pp.

Tucker,C.J.,Vanpraet,C.L.,Sharman,M.J.,Van Ittersum,G., 1985: Satellite Remote Sensing of Total Herbaceous Biomass Production in the Senegalese Sahel: 1980-1984. - Remote Sensing of Environment 17:233-249.

Wischmeier,W.H.,Smith,D.D.,1978: Predicting rainfall erosion losses - Agricultural Handbook 537. United States Department of Agriculture.

Appendix 1. Terms of Reference.

Background and description of the proposed inventory methods are given in Orgut-Swedforest 1984.

The team should be composed of professionals covering the following fields:

- Project Management
- Community Forestry
- Remote Sensing
- Environmental and land degradation monitoring
- GIS processing
- Program Specialist in GIS software development

The duties of the team are:

Assessment/inventory of two Priority Areas concerning:

- Community Forests distribution and woody biomass
- Land use , land cover
- Soil erosion and effect of soil conservation measures

Demonstration of the potential of a GIS for planning purposes on national, regional and local levels:

- Supply/demand of fuelwood
- Identification of problem areas on a priority scale for planning and implementation of land conservation measures, based on e.g. population distribution, crop production and livestock distribution.

The results of the study should be presented in the form of:

- Maps in the scale of 1:250 000, 1:50 000 and 1:25 000
- Area statistics
- Three dimensional landscape models
- Final report on the feasibility study (including evaluation of methodology accuracies)

Appendix 2. Satellite characteristics.

SATELLITE	LAUNCH	BANDS	PIXEL SIZE	SCENE COVER	PERIOD
Landsat-1	1972	MSS 4-7	56 x 79 m	185 x 185 km	18 days
Landsat-2	1975	MSS 4-7	56 x 79	185 x 185	18
Landsat-3	1978	MSS 4-7	56 x 79	185 x 185	18
Landsat-4	1982	MSS 1-4	56 x 79	185 x 185	16
		TM 1-5,7	30 x 30	185 x 185	16
		TM 6	120 x 120	185 x 185	16
Landsat-5	1984	MSS 1-4	56 x 79	185 x 185	16
		TM 1-5,7	30 x 30	185 x 185	16
		TM 6	120 x 120	185 x 185	16
Landsat-6	1988?	MSS 1-4	56 x 79	185 x 185	16
		TM 1-5,7	30 x 30	185 x 185	16
		TM 6	120 x 120	185 x 185	16
		TM 8?	15 x 15	185 x 185	16
Landsat-6	1992?	MSS 1-4	56 x 79	185 x 185	16
		TM 1-5,7	30 x 30	185 x 185	16
		TM 6	120 x 120	185 x 185	16
		TM 8?	15 x 15	185 x 185	16
Spot	1986	S1-S3	20 x 20	60 x 60	26
		Panc.	10 x 10	60 x 60	26
NOAA 6-10	1979-1986	AVHRR 1-5	1 and 4 km	2700 x 2700	18
Meteosat	1977	1-3	2.5 and 5 km	Western hemisphere, geo-stationary	

Approximate coverage cycle 1987

Landsat system	approx.	every 9 days or every 16 days
Spot -"-	- " -	- " - 3 days for non vertical coverage
Spot -"-		every 26 days for vertical coverage
NOAA -"-	- " -	- " - day
Meteosat -"-		every 30 minutes

MSS=Multispectral Scanning System, TM=Thematic Mapper, S1-S3= Spectral band 1-3, Panc.=Panchromatic